A Skeptical Biochemist

JOSEPH S. FRUTON

A Skeptical Biochemist

HARVARD UNIVERSITY PRESS
Cambridge, Massachusetts
London, England
1992

Printed in the United States of America
10 9 8 7 6 5 4 3 2 1

This book is printed on acid-free paper, and its binding
materials have been chosen for strength and durability.

Library of Congress Cataloging-in-Publication Data

Fruton, Joseph Stewart, 1912–
A skeptical biochemist / Joseph S. Fruton.
p. cm.
Includes bibliographical references and indexes.
ISBN 0-674-81077-5 (acid-free paper)
1. Biochemistry—History. I. Title.
QD415.F78 1992
574.19′2′09—dc20 91-29378
CIP

For Topsy,
in celebration of our
fifty-fifth anniversary

Contents

Preface

The title of this book resembles that of Robert Boyle's *The Sceptical Chymist*, which appeared in 1661, and that of Joseph Needham's *The Sceptical Biologist*, published in 1929. A prolific experimenter, Boyle described the composition of chemical substances in terms of empirically determined properties, and preferred "corpuscular" explanations of these properties to explanations in terms of the "substantial forms" and "real qualities" favored by the Aristotelian "Peripateticks" of his time. In his book of essays, Needham, a young member of the Biochemical Laboratory in Cambridge, defended what he called the "neo-mechanistic" attitude to biological problems, and cited some of the empirical physical-chemical knowledge of the 1920s to counter the "neo-vitalism" of leading contemporary biologists. Several years later, after his entry into the field of chemical embryology, Needham shifted his position toward "organicism."

In following the example of Boyle and Needham in my choice of title, I do not presume to do more than offer some skeptical thoughts about efforts to describe, advance, defend, or belittle the scientific endeavor to explain the properties of biological organisms through the study of the constitution and interactions of their chemical components. If the essays in this book deal with some of the questions raised by Boyle and Needham, these questions appear in a different form, as a consequence of the transformation of the conceptual framework of chemistry during the centuries which separated their books, and the transformation of the conceptual framework of the biochemical sciences during the decades which separate Needham's book from this one.[1]

After an introductory chapter, I offer some skeptical thoughts about recent discourse concerning what was once called the "scientific method," with special reference to the interplay of theory and practice in the biological, chemical, and biochemical sciences. In the third chapter I examine some features of the interaction of these areas of scientific inquiry since about 1800. The fourth chapter deals with problems in the study of the historical development of the chemical and biochemical sciences, and in the final chapter I consider some aspects of the evolution of the languages of these sciences. Although the perspective of each of these chapters is somewhat different, I hope that readers will find threads which link the philosophical, historical, and social aspects of the chemical and biological sciences to each other and to the growth of chemical and biological knowledge.

I claim no originality for the personal views presented in this book, for many of them reflect opinions expressed by earlier writers in the context of the scientific knowledge and social circumstances of their time, and I have reason to believe that views similar to mine are held by many of my colleagues in the biochemical sciences and in the history of science. There are numerous quotations, all in English; except where otherwise stated, the translations from French and German are my own, and may differ in detail from those published elsewhere. I am grateful to many authors and publishers for permission to quote or to translate copyrighted material.

Several individuals have given me valuable criticisms of early drafts; I am especially indebted to Sidney Altman, John T. Edsall, Frederic L. Holmes, Stephan Körner, Sofia Simmonds, and John Harley Warner for their help.

J.S.F.
New Haven, Connecticut
29 January 1991

1. Boyle (1661); for a complete entry, see Fulton (1961), p. 28. For Boyle's "corpuscular" philosophy, see Clericuzio (1990). See also Needham (1929, 1936).

A Skeptical Biochemist

CHAPTER ONE

Biochemistry and Skepticism

The human activity in which scientific hypotheses are devised, scientific observations and experiments are made, and the knowledge so gained is published in books and journals, is today a well-supported social enterprise. In most affluent nations, and some poorer ones, government agencies, private foundations, business organizations, research institutes, and educational establishments expend funds to promote this activity. The primary social purposes of present-day science in the wealthier nations are the development of new military weapons or new means of response to military attack, the search for solutions to current medical problems (with special attention to diseases prevalent in these nations), the invention of new industrial products, and the improvement of agricultural practice to increase the food supply. Although these, and other, social purposes are taken to justify the financial aid given to scientists, and often influence the choices scientists make in selecting problems for research, the *doing* of science is characterized by aims, methods, and values intrinsic to the scientific enterprise itself. Less public support is given to efforts to promote the *understanding* of these aims, methods, and values, of their relation to the social purposes of scientific research, and of the way in which both the intrinsic character and the public status of the sciences have been subject to historical change. In this book I offer my opinions about the efforts of some scientists, philosophers, historians, and sociologists to clarify problems which have arisen from the development of the sector of scientific inquiry that involves the interplay of chemical and biological thought and practice.

The emergence of the chemical sciences during the eighteenth and nineteenth centuries as the dominant scientific disciplines in the study of the special properties of the matter which constitutes the natural world came from the work and thought of diverse kinds of investigators, among whom there were not only natural philosophers but also craftsmen in the practical arts. Experimental skill in the isolation of supposedly individual chemical substances from natural sources (ores and oils, tissues and fluids of living things) and especially in the quantitative measurement of their properties and interactions, together with keen intellect in the formulation of hypotheses (many of which later fell into disfavor), produced a "mechanistic" science whose methodology and conceptual framework is still undergoing change. Moreover, by the early years of this century, the success of chemists in creating "unnatural" substances deemed to be of value in medicine, agriculture, industry, and warfare strengthened the public acceptance of the importance of chemistry in the hierarchy of the natural sciences.

In the biological sciences, with their historical ties to the practical arts of medicine and agriculture, the "mechanistic" view that the properties of living organisms could be explained in terms of the properties of their chemical constituents has been under attack to this day. By the end of the nineteenth century, the idea of a specific vital force *(Lebenskraft)* was no longer in favor, but the growing appreciation of the complexity of the intimate structure and physiological processes of living things gave rise to a series of "neovitalist" hypotheses, some of which invoked the existence of "living" chemical components (for example, "biogen") inaccessible to study by the available chemical methods. Even after specific hypotheses of this kind had fallen by the wayside, around 1930 the failure of disrupted biological systems to perform chemical processes (for example, the synthesis of urea) effected by surviving tissues was taken to mean that such processes are "linked to life." More important, advances in the study of physiological processes such as the transmission of hereditary characters from parents to progeny or the transformation of a fertilized egg into an adult organism led to an emphasis on "organicism," with a denial of the possibility that the properties of the whole could be explained in terms of the properties of its parts. This view has an honorable heritage, for Aristotle had written that "the formal nature is of greater importance than the material nature," and that the proper method of studying a physiological process is to determine its purpose. For example, he stated that "in dealing with respiration, we must show that it takes place for such or such a final object, and we must also show that this or that part of the process is necessitated by this or that stage of it."[1] This teleological or, in its recent formulation, "teleonomic" view of the functions of the parts of a living organism is a prominent feature of present-day thought among biologically-minded biochemists who use chemical concepts and methods to study physiological processes.

The structure and physiology of individual representatives of biological species have long occupied the attention of scientists of diverse philosophical inclination, and during this century many complex physiological processes previously thought to be inexplicable in chemical terms have been described in terms of the molecular structure of constituents of living organisms and of the interactions of such chemical substances with each other and with the chemical and physical components of their natural environment. These achievements have given the biochemical sciences, among which I include molecular biology, high public status, for reasons similar to those which led to the earlier recognition of the chemical sciences as central research disciplines. It must be recognized, however, that these important contributions to biological knowledge, with attendant consequences in efforts to control human disease or to improve agricultural crops, represent only initial and partial steps toward an understanding of the chemistry of life. Moreover, and most important, the biochemical approach has not constituted the whole of the endeavor to elucidate the properties of living organisms. The elaboration of Darwin's theory of natural selection, together with the growing knowledge of the process of the transmission of hereditary characters, has led to a cleavage between the areas denoted as "functional" and "evolutionary" biology. In the latter, explanation of the properties of biological organisms not only includes their genetic constitution, as modified by chance mutation, and its expression in their anatomy and physiology, but also emphasizes their interactions with their environment and the effect of

1. Aristotle, *De Partibus Animalium*, 640b, 642a.

these interactions on their reproductive capacity, with consequences for the survival and transformation of biological populations.[2]

As in the historical development of other areas of the collective human activity we now call scientific research, skepticism has played a large role in the interplay of chemical and biological thought. This ancient (and frequently unwelcome) mental attitude, ranging from faint doubt to outright disbelief, has been evident both *within* the individual areas of investigation as they separated into professional "disciplines" and among those scientists, philosophers, historians, and sociologists who have written *about* the sciences. Within the separate disciplines, as defined largely by the academic customs of the time, the participants have voiced skeptical views about the plausibility of theories offered by others and about the validity ("accuracy") of the observations or experimental data reported by colleagues in their own or a closely related field. Such skepticism, which often led to disputes among leaders in the profession, and to the dismissal of the claims of its less prominent members, was evident even during periods when there was a widespread consensus within a research discipline regarding its conceptual framework and the proper methods for gaining new empirical knowledge. In the modern chemical and biological sciences, at least since 1800, such periods of "normal science" have been relatively brief. Theories which had been for a time disregarded, or seemingly refuted, later won acceptance in the face of new empirical knowledge that was deemed to confirm a scientific claim.

On the other hand, when theories based on the methodology of one discipline were introduced into another area of inquiry, the skeptical attitude, and its attendant tensions, have persisted for many years. The introduction into chemistry of the "mechanistic" view that the explanation of natural phenomena was to be sought in the properties of "corpuscles" and of the physical forces which determine their interaction was challenged for at least two centuries by those who claimed that it was necessary to assume the existence of imponderable material substances or who insisted on the primacy of "energy" over "matter." In analogous fashion, the debate over the introduction of the ideas and methods of chemistry into biology has lasted for decades, and its resolution, if it ever comes, is not yet in sight. Chemists have been skeptical about chemical hypotheses of-

2. Mayr (1982).

fered by biologists who, in turn, have been skeptical about biological hypotheses offered by chemists. The resulting tensions have been transmitted through several generations to the present-day scientists who identify themselves as biochemists or molecular biologists. In contrast to the members of such research disciplines as organic chemistry, biochemists do not at present constitute a relatively homogeneous group with a consensus of opinion regarding the aims and conceptual structure of their area of endeavor. As a consequence of the historical development of the interplay of the chemical and biological sciences, some biochemists have been more chemically-minded in their choice of research problems and of the methods they use, and have tended to be less concerned than their biologically-minded colleagues with the physiological function of the biochemical constituents or processes chosen for study. Indeed, among the chemically-minded biochemists, some have followed research tracks which overlapped with those of the professional organic chemists of their time, whereas others have based their research on the concepts and methods of physical chemistry, or on the use of powerful new physical methods (X-ray crystallography, nuclear magnetic resonance spectroscopy, computer technology) in the study of the structure and interactions of biochemical substances. Similarly, among the biologically-minded biochemists, there are those who continued to advance knowledge of the physiological function of such classes of chemical substances as enzymes, hormones, immune bodies, neurotransmitters, and metabolic regulators or, as in the recent past, on the application of new methods to study the role of nucleic acids in a vast variety of biological processes. Other biochemists have sought to enter the arena of evolutionary biology, through chemical speculations about the "origin of life" or by comparison of biochemical processes or of the structure of individual biochemical substances in organisms occupying widely different positions in the time scale of biological evolution. Amid this diversity of scientific inclination, it has not been uncommon for leaders of one section of the biochemical community to disparage the efforts of those who have followed other research tracks.

The emergence of new specialties in the chemical, biological, and biochemical sciences, with claims for recognition of these specialties as distinctive areas of scientific inquiry and instruction, has led to some ambiguity in recent writings as to the meaning of the term "scientific discipline." I will consider this question more fully in a later chapter, and only note here that I consider a modern scientific discipline to represent a collective human activity in which natural objects, phenomena, or processes of a particular kind are subjected to systematic investigation, and in which the theoretical and empirical results are communicated to other participants in that specialized effort, to students, and to the scientific community at large. From this point of view, the most characteristic features of a scientific discipline are the nature of the research problems considered to fall within its scope, and the manner in which the conceptual framework and practical methodology are organized for the purpose of instruction. In the case of the biochemical sciences, the boundaries between separate areas of inquiry on particular kinds of research problems have changed (and sometimes disappeared) with the growth of chemical and biological knowledge, and the modes of instruction have reflected the transformations in theory and practice. Moreover, as a consequence of changes in the social and institutional status of particular specialties, at different times and in different places, different names were assigned to the same or very similar "disciplines."

Much has been written about the concepts and methods of modern scientific research by distinguished contributors to the growth of knowledge within their areas of scientific inquiry. For example, the book *Introduction à l'Étude de la Médecine Expérimentale*, by the nineteenth-century physiologist Claude Bernard, has been considered to be a classic exposition of the philosophy of research in the sciences allied to medicine. The list of eminent twentieth-century scientists who have offered their philosophical opinions about the "scientific method" is a long one, with a preponderance of physicists and biologists, and the variety of these opinions indicates the complexity of the problem of developing a unified theory of the scientific endeavor. Around 1840, the problem had been attacked by men such as Auguste Comte and William Whewell, but since that time the interplay of scientific thought and action has greatly accelerated, diversified, and been made more complex by many interrelated historical developments. Prominent among these developments have been the rise of large-scale industrial manufacturing, the invention of more rapid means of transportation and communication, increased financial support of scientific research in the wealthier nations, and changes in the institutional status of the sciences arising from the emergence of new universities, higher technical schools, and research institutes. One consequence of these developments has been an unprecedented rate of increase in the scientific population (slowed only by the 1914–1918 and 1939–1945 world wars), especially in research disciplines related to medicine, industry, and the military. Another was the formation of many relatively large research groups, initially in organic chemistry but later in such fields as experimental physics and the biochemical sciences, especially those now included among the so-called biomedical sciences. A third, and perhaps the most important, consequence was the proliferation of new instruments whose use opened unexpected vistas of scientific thought.[3]

During this century, many philosophers have attempted to improve on earlier descriptions of the "scientific method." They have emphasized the fictional nature of the view that theories arose in the minds of scientists from supposedly theory-free empirical knowledge. They also recognized that an understanding of the interplay of thought and action in scientific research required closer consideration of the historical changes in the social circumstances under which research was conducted, and of the attendant changes in the supposedly distinctive behavior of scientists in their search for new knowledge. To the logical analysis of the conceptual structure of scientific theories was added an examination of the historical development of the interplay of thought and action. This convergence of the efforts of some philosophers and of professional historians of the sciences has been paralleled by the studies of sociologists on the institutional and personal features of scientific research.

In the next chapter I offer some skeptical thoughts about recent philosophical discourse on the relation of observation and experimentation to the formulation of scientific hypotheses and about the criteria for assessing the value of such hypotheses. Although I will refer to the opinions of professional philosophers, I will not enter the arena of their disputations. This caution is not only a matter of prudence, but also one of skepticism regarding the extent to which modern analytical philosophy has illuminated the problems which have arisen from the interplay of the modern chemical and biological sciences. I will therefore consider primarily some of the writings of noted biologists, chemists, and biochemists about these philo-

3. Rothschuh (1976); Ackermann (1985); Gooding et al. (1989).

sophical problems. Although many of these writings reveal the influence of prominent philosophers, their more distinctive features are the tendency to reflect habits of thought and action developed during an earlier period of productive scientific work, the justification of the research styles of the authors, and the assertion of the autonomy of the particular area of scientific inquiry with which they wish to be identified. This book is no exception, and informed readers will find in it similar reflections of my own previous activity in biochemical research and education. They will also detect the influence of some of the critical writings concerning the relation of the history of philosophical thought to the development of scientific practice. In particular, I recognize that modern philosophers have made important contributions to the understanding of the structure of scientific theories and of their relationship to mathematical and especially to "common-sense" concepts.[4] It has also been a source of intellectual pleasure to find approaches which took account of the impact of the growth of scientific knowledge on the transformation of philosophical thought.[5]

Although many leading modern philosophers of science have recognized the problems inherent in past attempts to define the logical principles of an all-inclusive and time-independent "scientific method," in their discourse they have tended to exalt theory over practice, and to neglect the role of craftsmanship in the growth of scientific knowledge through the invention of new techniques and new instruments. This attitude is also reflected in the writings of some professional historians of science who consider their field to be primarily a part of "intellectual history." A major portion of this book is devoted to a discussion of the limitations of this approach in understanding the development of the interplay of the biological and chemical sciences. Whether the debate is about the notion that scientists should strive to find "simplicity" or "beauty" in their theories about natural objects and processes, or about the relative merits of "organicism" and "reductionism" in the search for explanations of biological phenomena, the important question, in my opinion, is not whether a theory is intellectually satisfying but

4. See especially Körner (1976).

5. In addition to the books cited in notes 3 and 4, those by Feyerabend (1975, 1987), Hanson (1958, 1971), Hempel (1965), Körner (1959, 1970, 1971), Kuhn (1977), Miller (1987), Nagel (1961), Schlesinger (1963), and Shapere (1984, 1987) have been most valuable.

whether it leads to new empirical discovery. Thus, if the experimental or observational test of a simple and beautiful theory leads to the discovery of what, for a time, may appear to be bewildering complexity, that theory, in my view, will have played a useful role in the development of an area of scientific inquiry. In the modern chemical and biological sciences, novel techniques and instruments have repeatedly proved to be decisive in attempts to unravel such complexity, and in the elaboration of older theories or their replacement by newer ones. There have also been what, from the perspective of present-day biochemical knowledge, might be considered as fads, fictions, and follies, and some of them will be mentioned later in this book. They deserve attention, for I believe it is necessary to recognize the limits at any time, including the present, of both rational thought and the empirical practice which inspires and guides such thought, in the continued exploration of the "chemistry of life." If the study of the history of this social activity has any value, surely one of its purposes ought to be to suggest that some of the scientific beliefs we now hold dearly may suffer the same fate as the now outmoded, though once highly regarded, theories of the past.

Since the Second World War, with the rapid growth of the scientific population and the increased involvement of governmental agencies in the support of scientific inquiry, much attention has been given to prominent psychological, social, and institutional features of the scientific life in recent times. Many of these studies have dealt with human qualities such as "creativity" or competition for recognition and rewards; others have stressed the role of scientists as leaders of research groups or university departments, or as entrepreneurs and politicians in the advancement of their scientific "territories." The public encouragement of such studies is not unrelated to matters of "science policy" in the organization of departments in universities and research institutes, and in the disbursal of funds in support of scientific research and education. It is regrettable, in my view, that with some notable exceptions, the writings of psychologists, sociologists, and social historians on these aspects of the scientific life suggest an ignorance of the detailed historical record of the development of theory and practice in the area of scientific inquiry under consideration. Indeed, in some cases there has been a disdain of that record on the ground that the principal aim of scientists is to promote their social status. It will be evident,

I hope, to readers of this book that I believe the social and institutional relationships among members of the chemical, biological, and biochemical communities to be intimately intertwined with the course of scientific investigation. To disregard the historical record of such investigation invites, in my opinion, the production of myths no less egregious than those of scientist-historians who have described the development of their specialties as an unbroken series of successes leading to the high ground of the present.

The final chapter in this book deals largely with the problems which have arisen from the changes in the languages of the biochemical sciences in response to new empirical discovery and from the transformation of the conceptual structure of these sciences since the eighteenth century. I believe it to be important to recognize the ways in which chemical substances and processes identified as components in the organized activity of biological systems have been classified and named at a given time, because ambiguities in the resultant languages reflect the stage of development of both the empirical knowledge and the accepted theories at that time. In the second part of that chapter I offer some comments about recent attempts to improve the literary quality of research papers in the biochemical sciences.

As a biochemist who has written several books and articles about biochemistry, I freely confess that I have participated in efforts to justify the existence and encouragement of our discipline. In 1951, my modest article on the place of general biochemistry in the university raised the hackles of the administrators of the Yale medical school. Later in the 1950s, Sofia Simmonds and I wrote a textbook whose aim was to emphasize the intimate relationship of biochemistry to modern chemistry and general physiology, and to counteract the tradition at many American universities of considering our field to be primarily an adjunct to medical research and practice. Subsequently, during the 1970s, I wrote a book about the historical development of the interplay of the concepts and methods of chemistry with those of biology, in the hope of counteracting the impression then gaining sway that a "revolution" had occurred in 1953 with the enunciation of the Watson-Crick model of DNA, and that a new science, "molecular biology," had been created. In the present volume, my purpose is still to emphasize the importance of chemistry in the study of biological problems and, in addition, to question the recent tendency among some philosophers and some philosophically-minded biologists and physicists to denigrate the efforts of several generations of chemically-minded biologists and biologically-minded chemists to bridge the gap between our knowledge of the properties of biological organisms and the chemical substances of which these organisms are composed. I hope that the skeptical thoughts offered in the succeeding chapters will help to define and perhaps clarify the nature of some of the differences in the recent discourse about the place of the biochemical sciences in the growth of our knowledge of the natural world.

I conclude with a statement by the noted anthropologist Clifford Geertz:

> The problem of the integration of cultural life becomes one of making it possible for people inhabiting different worlds to have a genuine, and reciprocal, impact upon one another. If it is true that insofar as there is a general consciousness it consists of the interplay of a disorderly crowd of not wholly commensurable visions, then the vitality of that consciousness depends on creating the conditions under which that interplay will occur. And for that, the first step is surely to accept the depth of the differences; the second to understand what the differences are; and the third to construct some sort of vocabulary in which they can be publicly formulated—one in which econometricians, epigraphers, cytochemists, and iconologists can give a credible account of themselves to one another.[6]

6. Geertz (1983), p. 161.

CHAPTER TWO

Perspectives on the "Scientific Method"

It is, I believe, fair to say that in the recent past few (if any) research workers in the biochemical sciences have sought guidance from the writings of contemporary philosophers in the choice of scientific problems, of rules of thought in the formulation of hypotheses, or of the procedures used to test the validity of the ideas generated during the course of the research.[1] For the actual practice of modern research on biochemical problems, therefore, the efforts of philosophers to develop a logic of scientific discovery, justification, and explanation appear to have been largely irrelevant. There can be no doubt, however, that the present conceptual frameworks of the biochemical sciences arose from the past interplay of philosophical thought and the ideas of those who sought to enlarge the scope of human knowledge and practice through observation and experimentation. This interplay, even if only dimly perceived by the scientific worker in the recollection of names such as those of Descartes and Lavoisier, or of Aristotle and Darwin, has left its imprint on the preconceptions and unstated assumptions which infuse the thought of the members of each particular area of research. It seems to me

1. In referring to the "biochemical sciences" I wish to emphasize the continuity of the interaction between many areas of chemical and biological investigation which have emerged as separate disciplines from what were once termed "natural philosophy" or "natural history" and from practical arts such as agriculture, medicine, pharmacy, or mining; see Fruton (1976). The term "discipline" has been defined in various ways. Except when otherwise stated (as in "institutional discipline"), my usage will refer to a specialized area of research and instruction, as well as to the values accepted in varying measure by those working, writing, and teaching within that area of scientific inquiry.

that even if the major trends in modern philosophy are judged to be irrelevant to successful scientific practice, the understanding of that practice is likely to be enriched by the critical examination of the historical development of the role of philosophical ideas in the growth of scientific knowledge.

From the centuries-old discourse on the proper methods in the investigation and explanation of the properties of natural objects and the phenomena they exhibit, philosophically-minded practitioners in the social activity we now call scientific research have inherited words intended to denote various theoretical aspects of the "scientific method." Among these words are *empiricism, rationalism, induction, deduction,* the *hypothetico-deductive method, positivism, phenomenalism, realism, pragmatism, verifiability, falsifiability, probability, simplicity, intuition, heuristic, hermeneutic.* Some pairs of these words have been apposed, as in empiricism and rationalism, or induction and deduction. During the course of the historical development of philosophical thought and of knowledge about the natural world, the meaning of these words (and of many others of a similar kind) has undergone change, and the words have been used in different ways by scientists who sought to describe the interplay of thought and action in their own successful research, and to offer advice about the proper methodology in their own discipline.

Although most of the modern literature (since about 1800) on the "scientific method" has come from professional philosophers and other non-scientists, many noted scientists also have expounded their views on this subject.[2] Among the chemists and biologists are Agnes Arber, Maurice Arthus, Claude Bernard, William Beveridge, Walter Cannon, Eugène Chevreul, Emil du Bois-Reymond, François Jacob, Henry Le Chatelier, Justus von Liebig, Ernst Mayr, Peter Medawar, Jacques Monod, Michael Polanyi, Jean Senebier, and Mathias Schleiden.

The Views of Peter Medawar

I begin with a discussion of the opinions offered by Medawar, because his elegant essays on the methodology of the sciences provide

2. For a bibliography of writings about the scientific method before 1900, see Laudan (1968). Hiebert (1987) has written a perceptive discussion of the philosophical writings of physicists.

some of the clearest recent statements by a scientist about his view of the relationship of analytical philosophy to scientific practice. Medawar attained fame through his research, with Rupert Billingham and Leslie Brent, on the problem of the rejection by one human patient of implanted tissues taken from another person (other than an identical twin). A noted experimental zoologist and a skilled histologist, Medawar was drawn into this research by the exigencies of the Second World War and the surgical problems encountered in the treatment of war wounds, especially severe burns. He interpreted his initial observations on the treatment of patients as indicating that the "homografts" were rejected by an immunological process. By 1944, after extensive work on experimental animals, he had devised reliable methods of transplanting skin, and had obtained results that strongly supported his hypothesis, which has become part of the conceptual framework of immunology.[3] Medawar had been interested in philosophy since his student days at Oxford, and after he received a Nobel Prize (with Macfarlane Burnet) in 1960, he wrote many essays on the philosophical aspects of scientific research. Most of these essays (often based on Medawar's public lectures) were collected in several books, one of which was entirely devoted to the problem of the role of induction and intuition in scientific thought.

In that book, one of the opening paragraphs contains the following statement, which offers a taste of Medawar's style of argument:

> The layman's interpretation of scientific practice contains two elements which seem to be unrelated and all but impossible to reconcile. In the one conception the scientist is a discoverer, an innovator, an adventurer into the domain of what is not yet known or not yet understood. Such a man must be speculative, surely, at least in the sense of being able to envisage what *might* happen or what could be true. In the other conception the scientist is a critical man, a skeptic, hard to satisfy; a questioner of received beliefs. Scientists (in this second view) are men of facts and not of fancies, and science is antithetical to, perhaps even an antidote to, imaginative activity in all its forms.[4]

3. Medawar (1986).
4. Medawar (1969), p. 2.

Later in the same chapter, Medawar states:

> If the purpose of scientific methodology is to prescribe or expound a system of enquiry or even a code of practice for scientific behavior, then scientists seem to be able to get on very well without it. Most scientists receive no tuition in scientific method, but those who have been instructed perform no better as scientists than those who have not. . . . Perhaps then we should no longer think of scientific methodology as a discipline of which the chief purpose is to teach scientists how to conduct their business, but rather as an attempt to get non-scientists to pull themselves together and smarten up and generally speaking be much more scientific than they are. Many modern methodological texts have therefore a strong orientation towards the social and behavioral sciences. . . . The "backwardness" of sociology (as in the nineteenth century of biology) has little now to do with a failure to use authenticated methods of scientific research in trying to solve its manifold problems. It is due above all else to the sheer complexity of those problems. I very much doubt whether a methodology based on the intellectual practices of physicists and biologists (supposing that methodology to be sound) would be of any great use to sociologists. On the contrary, the influence of inductivism . . . has in the main been mischievous. It has stirred up in some sociologists the ambition to ascertain the laws of social change, above all by the painstaking accumulation of data out of which general principles will in due course take shape.[5]

I have reproduced this excerpt not only because I share Medawar's skepticism about many of the generalizations offered by professional sociologists and social historians of the sciences (whose writings I will discuss in a later chapter), but also because I deplore his seeming failure to recognize that, in light of the complexity of the problems faced by any effort to examine and explain natural phenomena, among which, I believe, the development of human societies must be included, such an effort requires "the painstaking accumulation of data."

Medawar's criticism of the conventional description of the "in-

5. Ibid., pp. 8, 12–13.

ductive" method of scientific reasoning, as advocated by Francis Bacon and (later) by John Stuart Mill, is entirely persuasive if one accepts the rules of formal logic as being applicable to scientific practice. In his argument, Medawar echoes the following statement by Karl Popper:

> The best we can say of a hypothesis is that up to now it has been able to show its worth, and that it has been more successful than other hypotheses although, in principle, it can never be justified, verified or even shown to be probable. The appraisal of the hypothesis relies solely upon *deductive* consequences (predictions) which may be drawn from the hypothesis: *There is no need even to mention 'induction'.*[6]

Medawar acknowledges that "most original research begins with Baconian experimentation. We undertake to study a problem or a phenomenon for many different reasons; because it is interesting or important, because we have been led to it through earlier researches, or because we have been asked or told to do so; but whatever our motives may have been, the first thing to do is to find out what actually happens," but he then states that sciences which remain at a Baconian level of development "amount to little more than academic play."[7]

What is missing from Medawar's description of Baconian "inductivism" is an appreciation of the complexity of the historical interplay of philosophical thought and empirical practice, a deficiency that is also evident in Popper's *The Logic of Scientific Discovery.* If, in Popper's logic, it is necessary "to draw a clear line of demarcation between science and metaphysical ideas—even though these ideas may have furthered the advance of scientific thought throughout its history,"[8] a skeptical biochemist might venture the opinion that if one is genuinely interested in understanding the emergence and transformation of the conceptual frameworks of sciences such as chemistry and biology, it is necessary to consider whatever knowledge is available about the ideas, ancient and modern, which Popper may label as being "metaphysical," as well as the observations and

6. Popper (1959), p. 315 (italics in original text).
7. Medawar (1969), pp. 37–38.
8. Popper (1959), p. 39.

experiments (however these terms may have been defined by philosophers) made either by ancient (even prehistoric) peoples or by more recent investigators engaged in "academic play." Moreover, it seems to me that Medawar might have noted that the word *induction* has had wider meanings than the traditional dictionary one of "arguing from the particular to the general." Thus, the nineteenth-century philosopher Apelt wrote of "rational induction," based on Kantian philosophy, as distinct from the "empirical induction" attributed to Bacon and Mill.[9]

I believe, therefore, that Medawar's rejection of inductivism on the ground that it is an inadequate methodology of scientific research obscures the historical development of the various disciplines which we now consider to constitute the natural sciences. To this I must add my opinion that Medawar's advocacy of the "hypothetico-deductive" method of scientific reasoning also suggests a lack of historical perspective. According to Medawar, in scientific research "the generative act is the formation of a hypothesis" by non-logical intuition ("an imaginative preconception of what might be true"), and "we carry out experiments more often to discriminate between possibilities than to enlarge the stockpile of factual information." He adopts Popper's criterion of "falsifiability" as the proper method of testing the value of hypotheses, but allows for the possibility that "the act of falsification is not immune to human error." Although Medawar admits that "the most imaginative scientists are by no means the most effective; at their worst, uncensored, they are cranks," he claims that the "hypothetico-deductive system" gives "a reasonably lifelike picture of scientific enquiry."[10]

Claude Bernard and his Médecine Expérimentale

According to Medawar, "the wisest judgements on scientific method ever made by a working scientist were indeed those of a great

9. Ernst Friedrich Apelt, a mathematician, philosopher, and chemical manufacturer, was a disciple of Jakob Friedrich Fries, whose writings exerted an influence on the thought of the botanist Matthias Schleiden and of the physiologist Theodor Schwann. See Buchdahl (1973) for a valuable discussion of the views of Schleiden (1849) on scientific methodology.

10. Medawar (1969), p. 58. Moreover, Medawar failed to note that Popper's account of scientific theories ignores the difference between scientific information and explanation, as well as the difference between theories embedded in mathematics and those based on "common-sense" thought; see Körner (1986).

biologist, Claude Bernard,"[11] and it is therefore appropriate to note that the description in Bernard's *Introduction à l'Étude de la Médecine Expérimentale* of the sequence of ideas and experiments in the research that brought him fame differs markedly from that provided by the later careful examination of his laboratory notebooks. From the outstanding studies of Mirko Grmek and Frederic Holmes, it is clear that the ready acceptance of Bernard's published accounts of his work has introduced considerable mythology into the vast biographical literature about him.[12] Instead of the step-by-step use of keen observation, logical reasoning, and controlled experiment, advocated in the *Introduction,* Bernard stumbled, went down blind alleys, followed hunches suggested by the work of others, and grasped at explanations soon shown to be erroneous. To perceive that he shared these attributes with other, less successful, scientists may perhaps diminish the stature assigned by Medawar (and others) to Bernard's precepts of sound methodology, as well as to his opinions regarding the mechanist-vitalist controversy, but it does not lessen our appreciation of the importance of his experimental achievements in the development of experimental physiology. To see these achievements as the outcome of a prodigious capacity for assiduous effort and consummate manual skill in experimental surgery need not affect our regard for the acuity of his imagination in the formulation of scientific hypotheses, but we should also recognize that he was less disciplined in the use of such hypotheses than he chose to portray himself. Moreover, to focus too narrowly on the *Introduction* as the complete statement of Bernard's views regarding the place of inductive reasoning in scientific research overlooks some of the statements in his posthumous *Principes de Médecine Expérimentale:*

> In all the sciences, there are two distinct stages. They are: 1st, the stage of the science of observation; 2nd, the stage of experimentation. These two stages are necessarily and absolutely sub-

11. Medawar (1969), pp. 1–2. Medawar noted the relative neglect of Bernard's opinions in books on scientific methodology written in English. An exception was the book by Max Black (1954), pp. 14–20. It is also noteworthy that Popper (1972), p. 258, stated that Medawar had called his attention to Bernard's *Introduction.* Obviously, Bernard's philosophical writings figure largely in French books on the scientific method; see especially Bachelard (1938) and Canguilhem (1968, 1977).

12. Grmek (1973), Holmes (1974); see also Schiller (1967), Fruton (1979a).

> ordinate to each other. A science can never attain the stage of a science of experimentation without having passed through the stage of a science of observation. . . . Empirical observations are observations made without any preconceived idea and with the sole aim of establishing the fact without seeking to understand it. This kind of observation must always be the first basis of a science. . . . But once the facts of empirical observation are established, it is necessary to give them meaning, by deducing laws with the aid of hypotheses and observations, which are their touchstones, needed to verify them. It is to the latter observations that one should give the term *scientific observations*.[13]

Although Bernard noted that in some areas of scientific inquiry, such as astronomy, where initial observations "without a preconceived idea" may lead to ideas that guide further observations, without the possibility of "verifying" such ideas through controlled experimentation, his repeated emphasis was on the interplay of theory and skilled experimental practice. His definition of a *savant complet* was the following:

> 1° He determines a fact; 2° in connection with this fact, an idea arises in his mind; 3° in the light of this idea, he reasons, institutes an experiment, [and] imagines and realizes its material conditions; 4° From this experiment there result new phenomena which must be observed, and so on.[14]

In reading these statements, it is advisable to recall Bernard's bold assertion that

> empirical medicine is in total command today. *It is I who am founding experimental medicine, in its true scientific sense; that is my claim*. Consequently, in all that I will say, I will be obliged to 'polarize' myself toward *empirical medicine*, against which I battle and will always battle.[15]

13. Bernard (1947), pp. 2–3.
14. Bernard (1865), pp. 43–44.
15. Bernard (1947), p. 104 (italics in original text).

This egocentric attitude is evident in Bernard's cavalier treatment of the contributions of previous writers on many subjects and, in particular, on the methodology of the sciences. No mention is made of Whewell and Mill, and an even more glaring omission is the name of Jean Senebier, whose studies on the uptake, in the presence of sunlight, of carbon dioxide and the release of oxygen by green plants represent high points in the development of plant physiology. In 1802, Senebier published a three-volume work which includes notions about the relation of observations, theories, and experiments that are echoed in Bernard's *Introduction*.[16] If, as Grmek has suggested, Bernard had not read Senebier's book, perhaps some of Senebier's ideas were part of tacit knowledge among the biologists of Bernard's time. A similar historiographic problem, regarding the influence of Comte on Bernard's philosophical thought, has been ably discussed by Holmes.[17] As for Francis Bacon, Bernard had this (in part) to say:

> There is not much profit for a *savant* to discuss the definition of induction and deduction, nor the question of knowing whether one proceeds by means of one or the other of these so-called processes of the mind [*esprit*]. Nevertheless, Baconian induction has become famous and has been made the foundation of all scientific philosophy. . . . Bacon recognized the sterility of scholasticism; he understood and foresaw well the importance of experiment for the future of the sciences. . . . However, Bacon was not at all a *savant*, he did not understand at all the mechanism of the experimental method. . . . Bacon recommends that one should shun hypotheses and theories; we have seen nevertheless that they are the [essential] accessories of the method, as indispensable as the scaffolding for the construction of a house. . . . Moreover, I believe that the great experimenters appeared before the precepts of experimentation, just as the great orators preceded the treatises on rhetoric. . . . Consequently, it does not seem to me permissible to say, even in speaking of Bacon, that he invented the experimental

16. Senebier (1802); see Pilet (1962) and Marx (1974).
17. Holmes (1974), pp. 402–406.

method; the method so admirably practiced by Galileo and Torricelli, and which Bacon was never able to use.[18]

I have quoted these passages from Bernard's writings about scientific methodology in order to suggest that Medawar's inclusion of Bernard among the scientists who expounded the "hypothetico-deductive" method and who opposed Baconian induction raises more questions than it answers. It may be asked, for example, whether these writings were intended to represent a set of precepts Bernard followed in his own research, or an idealized picture of what he wanted his contemporaries to believe was the way he worked and thought in making his discoveries. Moreover, to what extent were these writings intended not only to reaffirm his public reputation, but also to promote the institutional status and autonomy of the discipline of which he claimed to be the founder?

Such questions, in my opinion, are not readily approached by means of the techniques of formal logic, whether of the Aristotelian or Popperian variety. Similar impertinent questions, not susceptible to critical analysis, may be asked about the extent to which Medawar's post-1960 writings about scientific research were influenced by his early exposure to "Oxford philosophy" and his later association with Karl Popper. I must leave it to future biographers and historians to consider such questions, and believe that it would be unfair to anticipate their judgment. To this I must add that my admiration of Medawar's experimental achievements was heightened upon rereading his major research reports on the problem of tissue transplantation.[19] Although written with the same elegance which characterized his essays, these reports convey, I believe, a more accurate and more stimulating picture of Medawar's scientific thought and practice.

Justus von Liebig on Francis Bacon

I turn next to another critic of Baconian inductivism, the celebrated nineteenth-century chemist Justus von Liebig. In 1863 he delivered

18. Bernard (1865), pp. 89–90.

19. Medawar (1944, 1945a). See also Silverstein (1989), pp. 285–291, and Mitchison (1990). Medawar's early interest in theoretical biology is indicated in his essay on size, shape, and age; see Medawar (1945b).

a lecture before the Bavarian Academy of Sciences on Bacon's place in the history of the sciences, with special attention to Bacon's method of induction. Liebig's principal criticism was that Bacon's glib acceptance of accumulated reports of observations, and his failure to recognize the importance of the work of men such as Galileo, Gilbert, or Hariot, led him to draw nonsensical inferences about natural phenomena. According to Liebig:

> The true method of scientific research [*Naturforschung*] excludes every arbitrary choice, and is diametrically opposed to that of Bacon. Every natural phenomenon, every process is always a whole, whose parts are unknown to our mind [*Sinne*]. We recognize the rusting of iron, the growth of a plant, but we know nothing about air, oxygen, or soil; our senses [*Sinne*] know nothing about all that goes on there . . . Our method does not rise from the simple to the complex, but we start with the whole, in order to discover its parts. . . . Bacon attaches great value in research to experiment; however, he knows nothing about its significance; he considers it to be a mechanical instrument which, when set into motion, produces a result; but in science all research is deductive and aprioristic; the experiment is only an aid to thought. . . . An empirical scientific research, in the ordinary sense, does not exist. An experiment that is not preceded by a theory, i.e. an idea, is related to scientific research in the same way that the shaking of a child's rattle is related to music.[20]

Liebig's convoluted German sentences and his idiosyncratic punctuation make his writing difficult to translate, but this quotation conveys a flavor of the polemical style that marked Liebig's writings between 1830 and 1850, when he was still active as the leader of a prominent chemical research group.[21] More to the point, Liebig's attitude toward the Baconian method, as expressed in 1863, is not the same as that in a letter he wrote in 1841 to Charles Gerhardt, whom Liebig considered to be one of his favorite pupils until Gerhardt presumed to question the accuracy of some elementary anal-

20. Liebig (1874), pp. 248–249.
21. Fruton (1988a); Holmes (1989a).

yses performed in Liebig's laboratory. In that letter, Liebig advised Gerhardt as follows:

> Do not give yourself up to any kind of theoretical speculations; they will serve to satisfy only the one person whose views you support, to gain you hundreds of enemies. Facts, particularly new facts, that is the only lasting merit; they speak more loudly, are appreciated by all minds, will bring you friends and win the respect of your adversaries.[22]

The "one person" was Auguste Laurent, who, with Gerhardt and other former *Liebigschüler*, notably Kekulé, through their "theoretical speculations" played major roles in the transformation of the conceptual framework of the organic chemistry of their time.

To return to Liebig's criticism of Baconian inductivism, I must add that the inaccuracy of Liebig's translation of sentences in the *Novum Organum* suggests some bias. One of these quotations is followed by Liebig's statement that "our method is Gilbert's method which Bacon condemns [*verdammt*], and therefore Bacon's method cannot be ours."[23] I will not quote Liebig's mistranslation of a part of Bacon's Aphorism LXIV, but ask instead that the reader compare it with the following English translation:

> The empiric school produces dogmas of a more deformed and monstrous nature than the sophistic or theoretic school; not being founded in the light of common notions (which, however poor and superficial, is yet in a manner universal, and of a general tendency), but in the confined obscurity of a few experiments. Hence, this species of philosophy appears probable and almost certain to those who are daily practised in such experiments, and thus have corrupted their imagination, but incredible and futile to others. We have a strong instance of this in the alchymists and their dogmas; it would be difficult to find another in this age, unless perhaps in the philosophy of Gilbert.[24]

22. Letter from Liebig, 27 June 1841 [Grimaux and Gerhardt (1900), pp. 53–54].
23. Liebig (1874), p. 250.
24. Bacon (1884), vol.3, p. 351.

It is not my intent to defend Bacon's philosophy, or to excuse his credulity, but to note that Bacon refers to Gilbert's philosophy and not to his experiments on magnetism. Moreover, the usage of the word "empiric" in Bacon's time was different from that during the nineteenth century.[25] Whatever faults may be found in Bacon's writings (and there were many), the fact remains that his message exerted great appeal, not only to the laity (as Liebig stated) but also to the natural philosophers who founded the Royal Society of London, as well as to many prominent scientists in nineteenth-century England.[26]

However one may judge the validity of Liebig's venomous condemnation of Bacon, there remains the question of his reasons for launching his attack in 1863, and for elaborating on the theme during the succeeding years.[27] All students of "Liebigiana" have been perplexed, at one time or another, by the vagaries of Liebig's mind and character. Can Liebig's anti-Baconian diatribes be an expression of his pique at the drop in his popularity among the English upper classes after Lawes and Gilbert had demonstrated the fallacies in his speculations and prescriptions of the 1840s regarding the improvement of soil fertility?[28] In a letter to Friedrich Wöhler (23 December 1866), Liebig affirmed that he was still "an admirer of this remarkable nation and truly love it."[29] Nevertheless, I find it likely that Liebig's writings about Bacon were a kind of sideswipe at the "gentlemen" (a term he used derisively) of English science whose adulation for him had cooled. Earlier in his career, when his principal scientific competitor was Jean Baptiste Dumas, it was France (where Liebig had learned chemistry from Gay-Lussac) that received most of his impassioned opprobrium. This nationalistic flavor of his polemical writings is also evident in his anti-Baconian writings. Thus, in his attack on Bacon, Liebig compared him unfavorably to the Germans Georg Bauer (Agricola) and Theophrastus von Hohenheim (Paracelsus).[30]

25. Williams (1985), pp. 115–117.

26. Although several earlier writers, notably Farrington (1953) and Ducasse (1960), have presented Bacon's ideas in a more favorable light, special attention should be given to the recent admirable book by Pérez-Ramos (1988).

27. Liebig (1874), pp. 255–295.

28. Fruton (1988a), p. 33.

29. Hofmann (1888), vol. 2, p. 227.

30. Liebig (1874), p. 241.

In his two final philosophical lectures before the Bavarian Academy of Sciences, Liebig treated the problems of the scientific method in more sober fashion, without the diatribes that marked his attack on Bacon. In these lectures, Liebig expounded a view of the role of both inductive and deductive reasoning that might be considered by present-day philosophers to approximate their description of the hypothetico-deductive method.[31] Moreover, as a former practicing chemist with an active interest in the social utility of his discipline, he recognized that scientific knowledge had emerged from the work and thought of people engaged in the practical arts. In my opinion, most modern philosophers of science do not appear to have stressed sufficiently this aspect of the development of the methodology of the chemical and biological sciences.

On Craftsmanship

It is common knowledge that, in the emergence of biology, keen observation of the properties of many kinds of living things and practical skill in the closer examination (as by dissection) of their anatomy, as well as the careful study of the fossil remains of extinct creatures, have guided the efforts to classify them and to formulate theories about their physiology and evolution. As in biology, the description and classification of the many kinds of non-living terrestrial objects, as well as technical skill in handling them, provided the starting points for the elaboration of the conceptual framework of chemistry. Moreover, as in biology, whose roots lie in practical successes and failures of primitive hunters, fishermen, farmers, and physicians, so also do the origins of chemistry lie in the experience gained by ancient craftsmen in the working of such materials as stone and metals, or in the preparation of articles of the diet such as bread and wine.[32] For the unlettered workers whose livelihood depended on the outcome of their practical efforts, the "explanation" of their observations by priests and philosophers in terms of theological belief or logical inference may have been less important than the economic circumstances of their lives, and many of them had learned from their own experience, and from the orally transmitted experience of their elders, that the trial-and-error method

31. Ibid., pp. 296–329.
32. Multhauf (1966).

was the proper way to promote their personal fortune and to advance the social status of their craft.

In the first century A.D., Pliny's *Natural History* provided a compendium of the observations made by practical men, and of the tools and methods they had devised. He included reports of many seemingly unlikely chemical phenomena, such as "all seawater is made smooth by oil, and so divers sprinkle oil on their face because it calms the rough element and carries light down with them."[33] Whether the observations recorded by Pliny were deemed to be fact or fiction, his book was widely read during the succeeding fifteen centuries and formed part of the empirical basis for the elaboration of a conceptual framework of chemistry drawn from various philosophical sources, especially the so-called Hermetic Books of that time, in which the ideas of Aristotle and other Greek thinkers were wedded to diverse religious traditions. It is customary to deride the thought underlying the practice of alchemy as one of the two great delusions of modern man, the other being astrology (which appears to have survived to this day even in some leading nations), but it cannot be doubted that the occult practice of many alchemists provided new knowledge about the properties of natural materials, as well as new or improved instruments (for example, distillation apparatus) for the purification of chemical materials. If we now judge the efforts of alchemists to transmute lead into gold or to isolate the "quintessence" which would cure all diseases to have been misguided attempts to gain fame and wealth, we must also admit that, in this fruitless and dangerous work, they also made observations that added much to the empirical groundwork of modern chemistry.[34] In the turbulent life of the sixteenth century, the cleavage between the empirical knowledge gained by chemical craftsmen and the philosophical ideas they had inherited became pronounced in the writings of Agricola and even in those of the Hermeticist Paracelsus and his followers.[35] Possibly, some philosophers may insist that the practical work and misguided thought of such craftsmen do not constitute a "scientific" activity, and that the value of all areas of inquiry into natural phenomena be judged in comparison with

33. Pliny (1938), pp. 359, 361 [book 2, chap. 106]; see Tanford (1980, 1989).

34. For the history of distillation, see Schelenz (1911) and Forbes (1948); for the idea of the "quintessence," see Taylor (1953).

35. Pagel (1982a, 1982b).

some idealized representation of the methodology of mathematical physics since the time of Galileo.

These thoughts raise several questions. Is the seeming neglect, in most of the present-day philosophical discourse about the "scientific method," of the contributions of skilled artisans a reflection of an attitude that places thinkers above workers, intellect above craft? It is understandable that such an attitude should be evident in the social philosophy of Aristotle, a member of the upper class in a slave society.[36] It is less clear, however, why leading modern philosophers have, for the most part, treated the rise of experimentation in sixteenth-century Europe as though it were solely an outcome of the thought of natural philosophers. Is it because laboratory work has been considered by many philosophers of science to be an unpleasant (though necessary) adjunct to the mental operations of a "true scientist," and something to be left to hired servants? Or is it because in the "logic of scientific discovery" there is no place for a technical chemist such as Johann Rudolph Glauber or even for those Victorian gentlemen who engaged in what some might now consider to have been "academic play" when they used a new instrument, the spectroscope of Bunsen and Kirchhoff, to examine the spectra of various natural pigments such as hemoglobin and chlorophyll?[37] Indeed, I believe that much could have been learned about the development of the "scientific method" from the informed and critical study of the manuals that have guided the practical work of successive generations of physicists, chemists, and biologists. Or does this reflect an attitude not uncommon among academic humanistic scholars, such as those in nineteenth- and early twentieth-century Oxford and Cambridge, toward the seeming fulfillment of Bacon's prophecy in the success of British industry, aided by imperial military and naval power?

It has been customary to make a distinction between "academic science," an activity said to be devoted primarily to research "for

36. Lloyd (1987). In their dictionary of biology, the Medawars (1983) reject the claim that Aristotle was a "man of science in the modern sense" on the ground that, as a biologist, he was only a collector and classifier of facts (as well as fictions). They also note that "no one did more than Bacon to bring the doctrinal tyranny of Aristotle to an end."

37. See James (1985). Among these Victorians were the Lucasian Professor of Mathematics at Cambridge, George Gabriel Stokes, the geologist Henry Clifton Sorby, and the physician Charles Alexander MacMunn.

its own sake," and technology (or applied science), in which the aim is to provide goods and services for social use.[38] This distinction is implicit, for example, in the present name of The International Union of Pure and Applied Chemistry. If Glauber was a technical chemist who sold his wares to apothecaries, and if later Heinrich Emanuel Merck was an apothecary who became a chemical manufacturer, both men contributed scientific knowledge that was used by "purer" chemists who followed them. Indeed, the history of twentieth-century organic chemistry cannot be written without taking account of the research achievements of chemists employed by the large-scale chemical firms that arose, first in Germany, then in Britain, Switzerland, and the United States. To be sure, many of these chemists might have preferred to become university professors but were denied such appointments for reasons, such as anti-Semitism, unrelated to their professional ability. These reflections may appear merely to endorse the now received view that the growth of scientific knowledge is intimately linked to the social circumstances in which research is conducted. I find it necessary to add, therefore, that vestiges of Aristotle's social philosophy are evident in much of the recent learned discourse on the "scientific method" with the tendency to exalt theory above practice, deduction above induction, and especially pundits above journeymen (however these terms may be defined).

In particular, I wish to emphasize the importance, in the development of the chemical and biochemical sciences, of work in which craftsmanship has been the most distinguishing attribute. One of the fundamental concepts of modern chemistry, since the eighteenth century, is the notion that one can extract from natural objects "pure substances" which are homogeneous and which have reproducible properties. I will return to the concept of chemical "purity" and now only call attention to the human qualities—among them consummate skill, unremitting effort, attention to detail, bold improvisation—evident in the work of the many unpretentious individuals who sought to isolate pure substances from natural materials. Such individuals might have been considered by some writers about the "scientific method" to be unworthy of inclusion in the category of "real scientists." To be sure, these qualities may be

38. See Ziman (1984). Keyser (1990) has discussed the role of the chemist Berthollet in the development of dyeing technology.

discerned in the work of many noted investigators whom such writers might be willing to admit to that category, but more attention needs to be given to the role that highly competent "journeymen" (often as research assistants of a less skilled person who headed a laboratory) have played in the growth of scientific knowledge.[39]

To give one example, out of the many that can be cited, I offer a sketch of the career of Moses Kunitz.[40] In 1913, when he was twenty-five years old, Kunitz became a full-time technical assistant of the noted biologist Jacques Loeb at the Rockefeller Institute for Medical Research in New York. During his association with Loeb, who recognized his promise and gave him much encouragement, Kunitz attended evening classes at Cooper Union and later at Columbia University, which granted him a Ph.D. (1924) in biological chemistry for a dissertation based on some of the work he had done in Loeb's laboratory on the interaction of proteins with salts. Loeb died in 1924 and was succeeded by John Howard Northrop, who invited Kunitz to remain in his laboratory and to continue work on the physical chemistry of proteins. In 1926 Northrop and Kunitz moved to the Princeton branch of the Rockefeller Institute, and soon afterward (perhaps spurred by James Sumner's report on the crystallization of urease) they embarked on efforts to crystallize the well-known protein-cleaving (proteolytic) enzymes pepsin and trypsin. It should be noted that by the turn of the century, several proteins had been prepared in crystalline form, largely in the laboratory of Franz Hofmeister (who had developed the "salting-out" method studied by Loeb and Kunitz), and most leading organic chemists (including Eduard Buchner and Emil Fischer) believed that enzymes are proteins, but that during the 1920s this view had lost general acceptance.[41] It was opposed, for different reasons, by two kinds of chemists and biochemists. Those who had embraced the new colloid chemistry, expounded by Herbert Freundlich and Wolfgang Ostwald, emphasized the role of the surfaces of colloidal protoplasmic particles in enzymatic catalysis. On the other hand, many organic chemists and biochemists followed the lead of Richard Willstätter in the belief that the active catalytic agents are small organic mol-

39. In many respects, but perhaps with a different emphasis, my comments about craftsmanship in scientific research echo the more extended discussion in Ravetz (1971).

40. A biographical memoir for Kunitz was prepared by Herriott (1989).

41. Fruton (1972), pp. 157–160.

ecules adsorbed on non-specific proteins. These notions gradually disappeared from the biochemical literature during the 1930s in the face of Northrop's crystallization of swine pepsin and his evidence for the view that the catalytic activity of this enzyme is a property of the protein. This success was followed by Kunitz's crystallization of a series of enzymes (and their inactive precursors) from bovine pancreas, and the achievements of Otto Warburg's associates in the crystallization of some of the enzymes involved in intracellular oxidations and fermentations.[42]

Not only did Kunitz crystallize the enzyme we now call trypsin (this name had been used in 1878 by Willy Kühne to denote the mixture of pancreatic enzymes he considered to be a single substance), but through a bold innovation he succeeded in extracting its precursor (trypsinogen), which he proceeded to crystallize. The bold step came from Kunitz's observation that trypsinogen is soluble and stable in cold dilute sulfuric acid, a treatment generally considered to be deleterious to "native" proteins. From such pancreatic extracts he isolated, in crystalline form, the precursor (chymotrypsinogen) of a new milk-clotting proteolytic enzyme (chymotrypsin), as well as a trypsin inhibitor and other pancreatic enzymes, notably ribonuclease and deoxyribonuclease. Moreover, during the Second World War Kunitz was asked to crystallize the enzyme hexokinase, present in yeast extracts, and he not only did so but on the way obtained other crystalline proteins, among them the enzyme pyrophosphatase.

A glimpse of Kunitz's style of craftsmanship is provided in the following account by Roger Herriott, one of his former colleagues in Northrop's laboratory. During the war,

> another laboratory had devoted considerable effort to the isolation of a plant protein [ricin] of great interest to the Department of Defense, but the investigator had been unable to crystallize the protein. A package of the material was sent to Kunitz with the request that he attempt to crystallize it. The package arrived late one afternoon. Kunitz dissolved some of the dry powder in water and placed small aliquots in a series of test tubes to which he added drops of dilute HCl, increasing the number of drops in each successive tube. A precipitate soon

42. Northrop et al. (1948); Warburg (1948). See also Carter (1990).

> began to appear in the middle of the series, and Kunitz held a turbid tube to the light from the window, remarking, after a few moments, "It looks granular." He placed the tubes in the refrigerator, and the next morning several tubes had crystals of what proved to be the active protein.[43]

The importance of Kunitz's contributions to the development of the conceptual framework of present-day biochemistry cannot, in my opinion, be overestimated. However, they have been overshadowed by the experimental knowledge gained from the intensive study (especially after the war) of the specificity, mechanisms of catalytic action, and intimate structure of crystalline enzymes. I would agree with those who may say that Kunitz was an exceptional figure among the "journeymen" of modern biochemistry only to the extent that his practical achievements were perhaps more significant in opening new avenues of research than those of many other skillful chemists and biochemists whose technical contributions were forgotten even by those who benefited from them. It is perhaps not surprising, therefore, that in recent discourse about the "scientific method," such lesser folk are frequently considered to have been only useful members of a successful research "team" whose leader reaped the accolades. To these observations I would add, without further comment, that Kunitz was elected to the National Academy of Sciences (U.S.A.) only in 1967, at the insistence of several members of its Section of Biochemistry, long after the election of those who gained such recognition through research on the crystalline enzymes he had provided.

With the emergence, during the nineteenth century, of relatively large research groups in the chemical institutes of German universities, the visibility of scientists noted for their technical skill began to diminish markedly. Although some heads of these laboratories were outstanding craftsmen (Robert Wilhelm Bunsen is a notable example), the success of such groups often depended on the participation of one or more assistants to whom students and other members of the group looked for guidance and practical instruction. The names of these highly skilled workers often appeared in print only in an acknowledgment at the end of a scientific article of which the

43. Herriott (1989), pp. 308–309.

leader of the research group was the principal author.[44] A similar development has been evident in the rapid growth, after the Second World War, of research groups in the biochemical sciences. Moreover, as during the first half of this century, when the specialized chemicals needed for research in organic chemistry became increasingly available from commercial firms, so also today investigators in the biochemical sciences can purchase preparations of hundreds of substances (enzymes, proteins, peptides, nucleic acids, polysaccharides, and so on) whose discovery and purification involved the craftsmanship of men and women who possessed some of the qualities exhibited by Kunitz. Clearly, as in other areas of scientific investigation, the role of skilled technicians in the chemical and biochemical sciences has changed with the development of the interplay of scientific theory and practice, with the introduction of new instruments or the improvement of older ones, and, more recently, with the invention of techniques for the automation of chemical processes and for the more rapid collection, calculation, and evaluation of numerical data.[45]

On Hypotheses in the Biochemical Sciences

I return to the views of some chemists and biologists regarding the mental operations in the "scientific method." Some form of Medawar's statement that, in scientific research, "the generative act is the formation of a hypothesis" may be found in numerous older writings of prominent scientists. For example, the noted organic chemist James Conant, who as president of Harvard initiated the publication of the celebrated *Harvard Case Histories in Experimental Science*, wrote in the introduction to the first volume of the series:

> Science we shall define as a series of concepts or conceptual schemes arising out of experiment and observation and leading to new experiments and new observations. . . . The cases presented in this series emphasize the prime importance of a broad working hypothesis that eventually becomes a new conceptual scheme. They illustrate the variety of mental processes by which the pioneers of science developed their new ideas. But

44. Fruton (1985).
45. See Shapin (1989).

> it must be emphasized that great working hypotheses have in the past often originated in the minds of these scientific pioneers as a result of mental processes that can best be described by such words as "inspired guess," "intuitive hunch," or "brilliant flash of imagination." The origins of the working hypothesis are to be found almost without exception in previous speculative ideas or in previously known observations or experimental results. Only rarely, however, do these broad working hypotheses seem to have been the product of a careful examination of all the facts and a logical analysis of various ways of formulating a new principle.[46]

To this quotation may be added another, from the valuable book *The Art of Scientific Investigation* by the pathologist William Beveridge:

> The hypothesis is the principal intellectual instrument in research. Its function is to indicate new experiments and observations and it therefore sometimes leads to discoveries even when not correct itself. We must resist the temptation to become too attached to our hypothesis, and strive to judge it objectively and modify it or discard it as soon as contrary evidence is brought to light. Vigilance is needed to prevent our observations and interpretations being biased in favour of the hypothesis. Suppositions can be used without being believed.[47]

An even more skeptical statement appeared in the preface to *De l'Anaphylaxie à l'Immunité*, by Maurice Arthus. After advising his readers to "seek facts and classify them—you will be the workers of science; imagine or accept theories—you will become their politicians," Arthus stated that the proper scientific method consists in "observing facts, in conceiving a hypothesis (a hypothesis is not a theory, it is simply a question), in judging the value of this hypothesis by experiment, in determining the meaning of this experiment, and in subjecting this interpretation to vigorous and tight criticism."[48]

46. Conant (1950), pp. 4–6.
47. Beveridge (1951), p. 52.
48. Arthus (1921), p. xv.

To this selection I add the matter-of-fact definition in the preface to a book entitled *The Scientific Method*, by Louis Fieser, another Harvard organic chemist:

> In attacking a problem of the usual type a scientist studiously looks for a clue or for a working hypothesis, carries out experiments planned to exploit the clue or to test the hypothesis, carefully observes and analyzes the results, and plans further experiments for advancement of the problem. Experiences cited in this book demonstrate that the scientific method is effective also in the solution of problems outside the domain of ordinary scientific research.[49]

Of the twenty-three chapters in Fieser's book, sixteen deal with the invention of weapons for the United States armed forces during the Second World War; among these weapons were napalm and various kinds of bombs. Little more is said about the "scientific method."

These quotations suggest something of the variety of opinion among chemists and experimental biologists about scientific hypotheses. Rarely, however, does one find the more exuberant view of Medawar, who, following Popper, stated that

> every discovery, any enlargement of the understanding begins as an imaginative preconception of what the truth might be. This imaginative preconception—a "hypothesis"—arises by a process as easy or as difficult to understand as any other creative act of mind; it is a brainwave, an inspired guess, the product of a blaze of insight.[50]

In his perceptive review of the book which contains this elegant passage, Max Perutz rightly noted that this description could not be applied to the work of Frederick Sanger, who determined the sequence of amino acids in bovine insulin, or to that of Dorothy Hodgkin, who determined the three-dimensional structure of insulin crystals:

> They started to explore the chemical formula and three-dimensional structure of insulin without any preconception: worse,

49. Fieser (1964), p. 5.
50. Medawar (1979), p. 84.

they didn't even have any clear idea of how they were going to find out what they wanted to know. Sanger did not work, à la Popper, by formulating hypotheses and then performing experiments to test them by falsification. Instead, he invented new chemical methods capable of solving problems that no one had even approached, since it was believed that they would defy solution.[51]

Sanger and Insulin: A Case History

As an admirer of Perutz, Sanger, and Hodgkin, both for their scientific achievements and for their personal qualities, I find it necessary to question the statement that Sanger and Hodgkin "didn't have any clear idea" of how they would proceed. I will only note my skepticism about the validity of that statement as it applies to the work of Hodgkin, but offer some extended comments about the circumstances under which Sanger undertook his work on insulin. First, I quote an excerpt from the reminiscences of Charles Chibnall, who succeeded Frederick Gowland Hopkins as head of the Biochemical Laboratory in Cambridge:

> I took over from Hopkins in August 1943, and was fortunate enough to be able to assemble almost at once a number of my old protein research workers from South Kensington—Kenneth Bailey, Williams, Rees, Tristram and MacPherson. Sitting in my office one morning I was turning over in my mind a suggestion I had made in my Bakerian lecture at the Royal Society the previous year, that the free amino groups in a protein over and above the epsilon-amino groups of lysine would be a measure of the number of polypeptide chains in the protein molecule. Abderhalden and Stix in 1923 (Hoppe-Seyl. Z. *129*, 143) used 2:4-dinitrochlorobenzene in an attempt to determine the terminal groups in a partial hydrolysate of silk fibroin, but had difficulty in separating the products concerned. The reagent, moreover, in $NaHCO_3$ solution would not react with amino acids unless heat was applied and this would in any case cause a partial breakdown of a protein. Recent research suggested to me that the problem was worth further investigation, for the reaction products 2:4-dinitrophenyl-amino-acids (DNP-amino-

51. Perutz (1989b), pp. 197–198.

acids) were bright yellow in solution, and hence amenable to the new partition chromatography of Gordon, Martin and Synge (1943, *Biochem.J. 37,* 39). In the past organic chemists had tended to steer clear of fluoroderivatives because of their toxicity, but the war gas research people at Porton were having them synthesized for this very reason, and [Rudolph] Peters had told me that they were often milder reagents than their chloro companions. I felt I should like to try out 2:4 dinitro-fluorobenzene to see if it would react with amino acids at room temperature. To obtain if possible further information about these fluorocompounds I decided to consult my friend McCombie, a lecturer in organic chemistry, whom I would occasionally meet for a glass of beer at a local tavern. When I told him my quest he at once roared with laughter and said that Saunders (a fellow lecturer in organic chemistry) had just made it!

On returning to the laboratory I decided to give the problem to Fred Sanger, who had just completed his Ph.D. under [Albert] Neuberger, who had just left Cambridge to join [Charles] Harington at the National Institute for Medical Research at Hampstead. I had known Neuberger for many years and had a high regard for his ability, so I guessed that Sanger had been well trained. I suggested to him that he should at once go and make love to Saunders, and try to get him to hand over a sample of his fluoro compound (2:4 dinitrofluorobenzene). He was successful, and as I had hoped it reacted under mild conditions without the need of heat. The resulting DNP-amino-acids could be readily separated and estimated by paper-chromatography. The next time I saw McCombie we agreed that a lot of good could come from a half pint of beer taken at the right time.

Sanger, at my suggestion, worked on insulin, the molecule of which appeared to be small, and as I expected made a fine job of his problem and had a draft of a paper ready in August 1945. My home life as a widower as well as university affairs were burdensome that year and I could give but scant time to the paper, but I realized that it made a significant advance and without hesitation I applied my South Kensington mode of publication—although I had suggested his problem he had made a good job of it, so I let him publish on his own.[52]

52. Chibnall (1987), pp. 31–32. In his reply to my request for his comments on

To this account I add a summary of Sanger's later work on the insulin problem, with some prefatory words about the "preconception of what the truth might be" in the case of a chemist who sought during the 1940s to determine the arrangement of the constituent atoms in the molecules of a substance under study.

Obviously, such a chemist had to base his work on the assumption that atoms and molecules (however these terms were defined since the time of Dalton) are indeed the sole constituents of natural materials, whatever some philosophers and some physical chemists may have written about the "metaphysical" character of the atomic-molecular theory. Next, the chemist had to convince himself and others that the sample of the substance constituted a homogeneous set of molecules, in other words that the sample was "pure." Throughout the development of modern organic chemistry, the most highly valued criterion of the purity of a chemical substance has been crystallinity. If a non-crystalline material could be converted by chemical means into crystalline derivatives, these derivatives provided the preferred starting point for the investigation of the constitution of the parent substance. This approach was eminently successful in the study of chemical compounds which had been obtained from biological sources and which, like the sugars, alkaloids, terpenes, or porphyrins, turned out to have molecular weights of less than about 1,000. Such crystalline compounds were recrystallized to assure the homogeneity of the sample, as judged by such criteria as constancy of data on their elementary composition, on their melting points, or on the optical activity of solutions of a portion of the sample. Clearly, the concept of "purity" depended on the selectivity and precision of the methods that were available at the time for examining the various properties considered to be characteristic of the individual substance under examination. As more selective and more precise methods and instruments (such as those based on the principles of chromatographic separation) were introduced into chemical practice, the possibility of detecting the presence of "impurities" was greatly increased. Occasional reminders are still needed, however, that "purity" (as applied to a chemical sample) is a postulated ideal state that may be approximated in

---

this passage, Dr. Sanger wrote that he "hadn't remembered it like that," and also noted that Chibnall "deserves more credit than he has been given." Clearly, the details of this episode in the history of protein chemistry merit closer scrutiny. See Synge and Williams (1990).

chemical practice, whereas heterogeneity (or the presence of an impurity) can be demonstrated by means of a sufficiently sensitive analytical procedure.

During the first decades of this century, leading organic chemists, among them Emil Fischer, declined to accept the experimental evidence for the view that proteins such as hemoglobin and egg albumin have molecular weights above 6,000. Also, leading biochemists, notably Hopkins, decried speculations about giant protoplasmic molecules and called for the more intensive study of the "intermediate metabolism . . . [of] simple substances undergoing comprehensible reactions," whereas S. P. L. Sørensen (1930) regarded many proteins as "reversibly dissociating systems." When, as a consequence of the work during the 1920s of Hermann Staudinger on the "macromolecular" constitution of rubber and the demonstration by The Svedberg that some highly purified proteins sediment in a strong centrifugal field as apparently homogeneous particles, it gradually became evident that proteins belong to a class of substances for which the words "molecule" and "particle" could have different meanings, depending on whether one defined these words in terms of the observed chemical or physical properties of the sample. Consequently, crystallinity, with its undoubted aesthetic appeal ("the beauty of crystals lies in the planeness of their faces") and its value in the study of relatively small organic molecules, lost some of its privileged status as a criterion of the purity of proteins, and other methods (for example, electrophoresis) were regarded as being more reliable.[53] It should be emphasized, however, that for biochemical studies of individual proteins and of other macromolecules of biological origin, crystallization still remains one of the most effective methods of purification, and is of course a necessity for the examination of their three-dimensional structure by means of the techniques of X-ray diffraction analysis. Moreover, apart from the technical skill required in the crystallization of proteins, the organization of molecules to form crystals has long attracted the attention of biologists who sought models for the organized complexity of the components of living systems. Some features of that search will be considered in the next chapter.

Insulin was crystallized in 1926 by the pharmacologist John Jacob Abel. When Sanger began his work on the constitution of this pro-

53. Pirie (1940); Shedlovsky (1943); Colvin et al. (1954); Vesterberg (1989).

tein, physical measurements on solutions of bovine insulin had shown that, depending on the nature of the solvent, its particle weight is either about 36,000 or 12,000, and it was assumed that the lower value represented its molecular weight. Apart from the assumption regarding the molecular size of insulin, the Chibnall group also held to the view that proteins are composed of chains of amino-acid units ("residues") connected by amide or imide linkages (-CO-NH- or -CO-N=) in which the alpha-carbonyl group (-CO-) of one amino acid is joined to the alpha-amino or alpha-imino group (-NH- or -N=) of another amino acid to form linear polypeptides. This "peptide" theory of protein structure was expounded in 1902 by Franz Hofmeister and Emil Fischer, and guided the research on proteins by their own groups, as well as those of other chemists, notably Thomas Burr Osborne and Edwin Joseph Cohn. During the 1920s and 1930s, however, the validity of this theory was questioned and various other proposals received attention. For example, there was a revival of interest in Fischer's speculation that diketopiperazines (cyclic compounds formed by the condensation of two amino acid units) might be structural elements of proteins, and the "cyclol" hypothesis offered by Dorothy Wrinch was endorsed by the noted physical chemist Irving Langmuir.[54] Some protein chemists were only ready to agree that "denatured" proteins (those formed by the treatment of "native" proteins that destroyed such properties as their enzymatic activity) are long chain polypeptides. Thus, after the demonstration that protein-cleaving enzymes such as pepsin, trypsin, and chymotrypsin can catalyze the hydrolysis of small peptides of known structure, Linderstrøm-Lang concluded from his studies on protein denaturation that his data

> provide sufficient basis for giving warning against the conclusion that genuine proteins contain peptide bonds because they are split by proteinases like trypsin. They give a certain indication that peptide bonds are formed or "appear" (like SH-groups) upon denaturation, but they are not conclusive enough

54. Fruton (1979b). During the 1930s many protein chemists (myself included) assumed that the specific properties of an individual "native" protein are closely related to the particular sequence, in the polypeptide chains of that protein, of the component amino acid units (-NH-CH(X)-CO-), which differ in the nature of their X groups. The decisive evidence for the validity of this view was only provided, however, during the 1950s by Sanger, through his work on the structure of insulin.

> to decide whether or not some hydrolysable peptide bonds are pre-formed in the molecules of the genuine globular proteins.[55]

Some present-day biochemists (or molecular biologists) who, in the face of present knowledge, accept without question the validity of the peptide theory of protein structure may find it difficult to understand these trends of the 1920s and 1930s, and may dismiss them as temporary aberrations. In my view, as a participant at that time in some aspects of the research on the problem, what was at issue during the 1930s was the question whether the constitution of proteins could be tackled profitably by means of the strategy organic chemists had used in the past to determine the structure of relatively small molecules.

After the purification of an organic chemical substance, this strategy, as developed through the interplay of theory and practice during the nineteenth century, involved first of all the quantitative determination of the composition of the substance (compound) in terms of its elements.[56] In the first three decades of that century, the type of elementary analysis for carbon and hydrogen introduced by Lavoisier began to be applied to a great variety of substances obtained from biological sources. The apparatus for the collection of the carbon dioxide and water produced upon combustion of a sample was successively modified and its reliability improved by such famous chemists as Gay-Lussac, Berzelius, and Liebig. In addition, methods were devised for the estimation of the content of other elements, notably nitrogen, sulfur, and phosphorus. By the 1830s, the accumulation of data on the elementary composition of organic substances was proceeding so rapidly that Liebig wrote: "Organic chemistry can now make one completely crazy. It seems to resemble a primeval forest in a tropical country, a vast thicket, without beginning or end, into which one may not dare to enter."[57] It should be added that the tasks assigned to the many students in Liebig's laboratory in Giessen between 1830 and 1852 largely involved elementary (or ash) analyses of natural materials, often in connection

55. Linderstrøm-Lang et al. (1938), p. 997.

56. See Freund (1904). For the development of organic analysis, see Dennstedt (1899), Holmes (1963), and Hickel (1979).

57. Hofmann (1888), vol. 1, p. 604.

with his disputes with other chemists about the validity of their analytical data.[58]

Those who had been introduced to organic chemistry during the 1920s and 1930s took for granted the consequences of the insight of Dalton in recognizing the importance of the relative weights of atoms, that of Berzelius in devising a new symbolic language of chemistry, that of Gerhardt in developing the theory of "types" of atomic groupings, that of Kekulé in advancing a theory of "valence" and the idea of chemical "structure," that of van't Hoff and Le Bel in visualizing the spatial arrangement of atoms in organic compounds, that of Rutherford and Bohr in formulating a planetary theory of the nature of atoms, and that of Lewis in the development of the electronic theory of chemical bonding. Also, several generations of chemists had developed a large variety of techniques for the selective cleavage of organic compounds to substances of known constitution, and in many cases the structure of a compound was deduced from overlapping features of the resulting fragments. Such hypotheses were then tested in attempts to effect the unambiguous synthesis of the compound under study. All these (and other) ideas and techniques, as modified by later empirical discovery and theoretical interpretation, have been fundamental in the development of the conceptual framework of modern biochemistry. This seemingly "Whiggish" account may raise the hackles of historians of science who have studied the uncertainties and disputes which attended the development of nineteenth-century organic chemistry. My point is that this account applies equally to the relatively slow growth of knowledge about proteins as compared with the rapid elucidation, during the period 1870–1945, of the structure of smaller molecules of biological origin. Although the readers of the textbooks of the 1920s and 1930s, in which these ideas and techniques had been enshrined, may not have known the tortuous history of nineteenth-century organic chemistry, such students had acquired a set of preconceptions which influenced their later approach to the study of the chemical structure of the more complex biological macromolecules. I have discussed elsewhere some of the features of the development, during the years 1870 to 1945, of ideas about the chemical

58. Morrell (1972); Fruton (1988a). Liebig's students did not engage to a significant extent in organic synthesis; see Brooke (1971).

constitution of the proteins, and will note here only the following points.

1. Although a few amino acids had been isolated from protein hydrolysates obtained by treatment with acids or from natural sources (for example, urinary calculi) before 1870, it was only by about 1910, after the work of agricultural chemists such as Heinrich Ritthausen and Thomas Burr Osborne, of physiological chemists such as Albrecht Kossel and Franz Hofmeister, and of organic chemists such as Emil Fischer, that it came to be fully appreciated that nearly all the then-known proteins were composed of about eighteen different amino acids whose individual structure had to be determined by the available methods of organic chemistry (including synthesis), and that procedures were needed for the separation of these amino acids from the mixtures formed upon the hydrolytic cleavage of proteins and for the quantitative estimation of the proportion of each amino acid in such mixtures. The determination of the chemical structure of the individual amino acids considered in 1910 to be "elementary" units of proteins, and the few additional ones identified later, was completed relatively quickly, but the intensive efforts of a succession of noted chemists (among them Donald Van Slyke, Max Bergmann, and Charles Chibnall) to devise reliable analytical methods provided only partial solutions to the problem.[59]

2. During the period 1840 to 1900, many physiologists and physiological chemists (among them Claude Bernard, Ernst Brücke, Felix Hoppe-Seyler, and Willy Kühne) studied the digestion of dietary proteins by "ferments" (Kühne termed them "enzymes") such as pepsin and pancreatin, and described the process as the conversion of proteins, via "albumoses," to "peptones." Relatively little attention was paid to the appearance of small amounts of the amino acids tyrosine and leucine which, because of their sparing solubility, often crystallized from the test solutions. The products of greatest interest were the peptones, which, because of their ability to pass through membranes, were thought to represent the intermediates in the metabolic conversion of dietary proteins into blood proteins, for it seemed "wasteful" for an animal organism to carry the breakdown of dietary proteins all the way to their constituent amino acids. Some physiologists, notably Kühne and his associates Russell Henry Chit-

59. Tristram (1949).

tenden and Richard Neumeister, expended much effort to purify and characterize the albumoses and peptones, and believed that they had obtained several of them in a "pure" state. With the widespread acceptance, after 1900, of the peptide theory of protein structure, the albumoses and peptones were termed "polypeptides," and Emil Fischer claimed that his research group had synthesized peptides which possessed the properties of the albumoses. However, for the study of the constitution of the proteins, the chemical methods then tested for their utility in the separation of polypeptides were entirely inadequate. Although, in 1908, Hofmeister reiterated his conviction that the isolation of well-defined cleavage products was the most important requirement for the determination of the structure of proteins, he also noted that "their enzymatic breakdown always leads to mixtures of many substances of unequal size . . . and the isolation of chemically-characterized albumoses or peptones is a thankless task . . . as a consequence of the size of the protein molecule."[60]

I suggest, therefore, that the attention given during the 1920s and 1930s to alternative theories of protein structure was in part a consequence of the fact that although the strategy based on acceptance of the peptide theory had been formulated, the chemists who held to that theory did not have general methods for the quantitative determination of the amino acid composition of proteins or for the separation of the products of partial hydrolysis by enzymes or dilute acids. In retrospect, it is clear that the turnabout came with the introduction into protein chemistry of a general separation method whose origins lay in the middle of the nineteenth century and which, in 1906, Michael Tsvet named the "chromatographic method" because he had used it to separate natural pigments. Partly in response to an ill-fated hypothesis of protein structure, offered by Bergmann and Niemann in 1938, Archer Martin and Richard Synge were led to develop, during the 1940s, a chromatographic procedure that proved to be the forerunner of the many more elegant and more precise methods and instruments available today.

When Sanger undertook the study of the arrangement of the amino acid units of bovine insulin, although he could not have had a "preconception of what the truth might be," it is hardly credible that he "didn't even have any clear idea of how [he was] going to

60. Hofmeister (1908), p. 276. See also Siegfried (1916).

find out what [he] wanted to know." First, it should be noted that he made two assumptions: (1) insulin is a relatively small protein, and therefore more likely to yield to attack than a larger molecule, such as hemoglobin; (2) insulin is composed only of amino acid units joined by peptide bonds to form linear polypeptides, each of which was expected to have a free alpha-amino group at one end of the chain and an alpha-carboxyl group at the other end. Also, at that time, members of Chibnall's research group (MacPherson, Rees, Tristram) had nearly completed their determination of the composition of the mixtures formed upon the hydrolytic cleavage of the peptide bonds of insulin. This task was a formidable one, beset with many pitfalls, but the results accounted for about 96 percent of the expected total, and showed that per unit of 12,000, insulin is composed of about 104 amino acid units of seventeen different kinds. Moreover, Sanger was able to use (and improve) the chromatographic procedures devised by Martin and Synge for the separation of amino acids and peptides. It does not, in my view, diminish the magnitude of Sanger's achievement to acknowledge (as he did in his research papers) the importance of these parallel contributions of his colleagues in guiding his approach to the insulin problem. On the contrary, I would stress the fact that, after about eight years of unremitting experimental effort, Sanger and his associates (notably Porter, Ryle, Thompson, and Tuppy) succeeded in providing a two-dimensional structural diagram of the insulin molecule whose validity was repeatedly confirmed afterward (including by eventual chemical synthesis); this achievement reflects their unswerving commitment, keen insight, remarkable skill, meticulous care, and disciplined self-criticism in the interpretation of their data.

I can offer here only a brief sketch of the route followed by Sanger in his work on insulin. By means of the DNP-technique which he developed, he established the presence of two amino-terminal amino acids—glycine and phenylalanine (the latter had been identified by previous investigators)—in amounts indicating the existence of four polypeptide chains per unit of 12,000. Other chemists, using different chemical techniques, also identified alanine and asparagine as carboxyl-terminal amino acids. As Sanger's work proceeded, he recognized that there were only two kinds of chains, which he labeled A and B, suggesting that the minimum molecular weight of beef insulin is about 6,000, a value supported by physical measurements. He also showed that the A and B chains are joined by disulfide (-S-S-) groups in two of the three cystine units in the molecule. The A and B chains were then separated by mild oxidation of the disulfide groups, and the amino acid sequence in each of the chains was determined by skillful use of chromatographic separation and amino acid analysis of the peptide fragments formed upon partial hydrolysis by means of dilute acid or by enzymes such as trypsin, chymotrypsin, or pepsin. Many overlapping pieces were obtained, and from their individual compositions and end groups, the sequence of the two chains and the location of the disulfide bridges were then deduced.[61] In the decades that followed, Sanger's procedure was replaced by more rapid methods which required less chemical skill, but it should also be noted that, in contrast to the sequence Sanger assigned to bovine insulin, many of the sequences initially reported for some proteins (for example, ribonuclease A and papain) were found to require correction. By now, hundreds of amino acid sequences have appeared in the biochemical literature, and I believe it is safe to predict that, if subjected to critical experimental test, some of them will also be found to have been inaccurate.

I have dwelt at some length on Sanger's work, and on some of what preceded it, in the hope that some professional philosophers of science or some philosophically minded physicists and biologists may be impelled to revise their descriptions of the "scientific method" and their precepts for its proper use by taking account of the way in which success has been achieved in the solution of problems in the area of scientific inquiry that overlaps the spheres of chemical and biological interest. Some of these people may consider the example I have chosen to have been only a manifestation of the kind of "normal" science from which no new "paradigm" had emerged, although I doubt whether Thomas Kuhn, with whose name these terms have been associated in the recent past, would regard it so. Some present-day professional historians of science may describe my choice of Sanger's achievement as another instance of the kind of "Whiggism" which they consider (sometimes rightly) to afflict scientists who presume to enter into their domain. Whatever merit such criticisms may have, I can only point to what I believe to be a historical fact: that, after Sanger's work on insulin, the further course of research on the structure of proteins was no longer encum-

61. Sanger (1952); Ryle et al. (1955). For a retrospective account, see Sanger (1988).

bered by arguments about the kinds of hypotheses offered during the 1920s and 1930s. Moreover, I suggest that the informed examination of other significant episodes in the growth of biochemical knowledge would lead to a similar conclusion. Regrettably, the many tomes and articles about the "scientific method" in the biochemical sciences have largely neglected the place of organic chemical practice in the rise and fall of competing theories. This neglect may perhaps be a consequence of ignorance of or indifference to the conceptual structure and methodology of modern organic chemistry. Another possible reason is the disposition to relegate to the status of "journeymen" outstanding chemists who did not participate in debates about theories.

The Perils of the Search for Simplicity

The two-dimensional diagram which Sanger presented for the arrangement of the amino acid units of insulin might be described as the statement of a rather complex theory because some chemists or philosophers who seek unity and simplicity in our understanding of the natural world might have found it to be more satisfying if there were more regularity in the pattern of the linear structure of the two polypeptide chains of the protein. This possibility invites some thoughts about the much-discussed aspect of the "scientific method" dealing with the criterion of simplicity in the evaluation of the relative merits of scientific theories.[62] Thus, homage has been paid to the principle ascribed to William of Ockham (often spelled Occam), the great medieval theologian and philosopher, who criticized the Scholastic dogma of his time by insisting that one must not multiply entities beyond necessity. His doctrine (or "razor") has been invoked repeatedly in succeeding centuries, up to the recent past, on the ground that simpler hypotheses, involving the fewest assumptions, are more useful than more complex hypotheses. Among modern biochemists, the principle of Ockham's razor has been cited largely by those inclined toward physico-chemical approaches to experimental problems.[63] It should be noted, however, that the simplicity required by Ockham's razor is rather different

62. See Bunge (1961), Sober (1975), and Miller (1987), pp. 245–262.
63. For example, see Northrop (1961).

from the simplicity sought through geometric or other mathematical representations.[64]

In the search for simplicity, clarity, and elegance in the formulation of biochemical hypotheses, modern scientists have used analogies (for example, Emil Fischer's lock-and-key analogy in describing the specificity of enzyme action), or two-dimensional diagrams (for example, the depictions of metabolic pathways, such as Krebs's ornithine cycle of urea synthesis), or three-dimensional models (for example, the Watson-Crick double-helical model of DNA). These three examples have remained in our memory because they stimulated fruitful research effort, and have been transmitted in biochemical textbooks, while many other such analogies, diagrams, and models have been forgotten and consigned to the dustbins of the history of science. These discarded biochemical hypotheses include some of the ideas that were, for a time, hailed because they were considered to be simple and elegant descriptions of natural objects and phenomena in the language of arithmetic or geometry and also, no doubt, because of the unquenchable hope that, in light of the seemingly increasing complexity of the observable natural world, one might discern, through the use of mathematics, a "logic" that provided evidence for the assumption that, after all, this world was really less complex than it seemed to be.

The search for simplicity in the explanation of natural phenomena through the use of a theory of numbers began as early as 600 B.C., with the efforts of Pythagoras of Samos and his disciples to develop numerical hypotheses about music and astronomy. The most beautiful harmonies produced by the lyre and the flute were explained in terms of ratios or combinations of numbers; so also were the previous extensive observations (over a period of centuries) by Babylonian sky-gazers who found periodicity in the displacement of the heavenly bodies. Aristotle found special virtue in the number seven. It is beyond the scope of this chapter to trace the later history of this ancient faith in numerology except to note that the emergence, during the seventeenth century, of quantitative experimental measurement appears to have reflected this confidence in the power of numbers to explain natural phenomena. Like other methods in the practice of scientific research, however, the value of quantitative measurement in the formulation and test of chemical and biological

64. I am indebted to Stephan Körner for calling my attention to this distinction.

hypotheses depends on the nature of the objects or processes under study and on the precision of the analytical procedures and instruments that are used. One of the best-known examples from the history of nineteenth-century chemistry is provided in the report, on 31 December 1808, by the outstanding experimental chemist Gay-Lussac that "compounds of gaseous substances with each other are always formed in very simple ratios, so that representing one of the terms by unity, the other is 1, or 2, or at the most 3."[65] Gay-Lussac also noted that his extensive data were "favorable to Dalton's ingenious idea that combinations are formed from atom to atom." Although Dalton did not accept Gay-Lussac's "law of combining volumes," the connection between it and the atomic theory was soon appreciated and fruitfully applied by the chemist Berzelius, whereas others (for example, the physicist Ampère) used Dalton's hypothesis to advocate the mathematical idea of "point atoms." Later, numerology played a role in the formulation of many chemical hypotheses, such as those relating to isomerism.[66] On many occasions, however, in striving for simplicity and elegance, chemists and biochemists have been led astray by their penchant for numerology because they allowed imagination and intuition to take precedence over attention to the limits of their experimental procedures and the validity of their empirical data. The history of protein chemistry during the 1920s and 1930s offers some examples.

During the 1920s, The Svedberg constructed (and later improved) an "ultracentrifuge" which permitted optical observation of the sedimentation of colloidal particles.[67] Contrary to his expectation, hemoglobin behaved as a homogeneous substance with an apparent particle weight of about 68,000, and subsequently many other protein preparations (among them myoglobin, casein, gelatin, and hemocyanins) were also found to sediment, under some conditions, as "monodisperse" sets of particles with apparent molecular weights ranging from about 17,600 to about 5,000,000. Although earlier estimates, notably those based on the careful measurements by Gilbert Adair of the osmotic pressure of protein solutions, had given values in the range of 34,000 to 74,000, most organic chemists continued to endorse Emil Fischer's pronouncement that such large particles are

65. Gay-Lussac (1809), p. 234.
66. Rocke (1984), pp. 22–47; Rouvray (1975).
67. Claesson and Pedersen (1972).

aggregates of units smaller than about 6,000. In one of his first papers on proteins, Svedberg introduced numerology into the interpretation of his data:

> The proteins with molecular weights ranging from about 35,000 to 210,000 can, with regard to molecular weight, be divided into four sub-groups. The molecular mass, size and shape are about the same for all proteins within such a sub-group. The molecular masses characteristic of the three higher sub-groups are—as a first approximation—derived from the molecular mass of the first sub-group by multiplying by the integers *two, three* and *six*.[68]

A few years later, after Svedberg had found that some proteins had an apparent molecular weight of about 17,600, the series of multiples became 2, 4, 8, 16, 24, 48, 96, 192, and 384, with the hemocyanins in the largest sub-group. He then stated that

> not only the molecular weights of the hemocyanins but also the mass of most protein molecules—even those belonging to chemically different substances—show a similar relationship. This remarkable regularity points to a common plan for the building up of the protein molecules. Certain amino acids may be exchanged for others, and this may cause slight deviations from the rule of simple multiples, but on the whole only a very limited number of masses seems to be possible. Probably the protein molecule is built up by successive aggregation of definite units, but that only a few aggregates are stable. The higher the molecular weight the fewer are the possibilities of stable aggregation. The steps between the existing molecules therefore become larger and larger as the weight increases.[69]

Svedberg had been awarded the Nobel Prize in Chemistry in 1926, and his attempt to introduce simplicity and unity into protein chemistry was treated with appropriate respect; during the 1930s skepticism about the validity of his numerical hypothesis was muted.[70]

68. Svedberg (1929), p. 871.
69. Svedberg (1937), p. 1061.
70. Pirie (1940, 1962).

Indeed, Svedberg's hypothesis encouraged Max Bergmann, and his associate Carl Niemann, to develop further the idea of a numerical regularity in the structure of proteins. On the assumption that a molecular weight of about 35,000 represents 288 amino acid units (whose average residue weight was taken to be about 120), Bergmann and Niemann concluded that the total number of amino acid residues in proteins of higher molecular weight falls into a series of multiples of 288. Moreover, on the basis of analytical data on the amino acid composition of acid hydrolysates of several protein preparations, they offered a periodicity hypothesis of the sequence of amino acid units in the polypeptide chains of proteins. This hypothesis stated not only that the content of each kind of amino acid unit in a protein can be denoted as a multiple of two and three ($2^n \times 3^m$) but also that each kind of amino acid recurs at a regular periodic interval in the polypeptide chains of proteins.[71] Periodicity hypotheses of this kind had been offered before. In 1904 Albrecht Kossel concluded from analytical data on the amino acid composition of the protamine clupeine that the amino acid arginine (Arg) recurs regularly in this protein in the repeating triplet -Arg-Arg-X-; in 1934 William Astbury inferred from the available data for the amino acid composition of gelatin that every third residue could be a glycine unit and every ninth a hydroxyproline unit; and in adding his values for the proline content of this protein, in 1935 Bergmann interpreted the available data for gelatin as supporting "the suggestion that these three amino acids occupy a definite periodic arrangement within the protein."[72] The Bergmann-Niemann hypothesis, therefore, was a generalization to all proteins of these earlier specific speculations.

To this summary of the Bergmann-Niemann hypothesis I add a personal note. In 1934, immediately after receiving my Ph.D. at Columbia University, I had the good fortune to begin work at the Rockefeller Institute in the newly established laboratory of Max Bergmann, a distinguished organic chemist who had been the last principal assistant of Emil Fischer, and who had made notable contributions in both carbohydrate and protein chemistry. Upon the Nazi takeover in Germany, Bergmann was obliged to leave his post

71. Bergmann and Niemann (1937).

72. Bergmann (1935). For Bergmann's early interest in numerology, see Fischer (1916).

as Director of the Kaiser-Wilhelm Institute for Leather Research in Dresden. Two years earlier, he and his associate Leonidas Zervas had published their famous paper on the so-called carbobenzoxy method of peptide synthesis, which opened a new stage in the development of this area of protein chemistry.[73] Soon after I joined Bergmann's group, Zervas came to New York, and because we shared a laboratory, he became my mentor. Whatever skill I acquired in the art of peptide synthesis came from my association with this outstanding chemist. In the years that followed, until the Bergmann group was obliged in 1942 to contribute to the war effort, my research dealt largely with the specificity of proteolytic enzymes, and the most rewarding results came from the finding of synthetic peptide substrates of known structure for well-defined proteinases such as pepsin, trypsin, and chymotrypsin.[74] When Zervas returned soon afterward to his native Greece, where he established an important school of peptide chemistry, his place in the laboratory in which I worked was taken by Carl Niemann. At that time, Bergmann's personal research effort was devoted entirely to the search for specific reagents for the amino acid analysis of protein hydrolysates, and Niemann's assignment was to use these and other available methods to determine the amino acid composition of several protein preparations (silk fibroin, gelatin, bovine hemoglobin, ovalbumin, bovine fibrin). As a daily observer of the progress of Niemann's work, I found it difficult to accept the hypothesis which emerged from his data on the ground that adequate evidence was not at hand for the homogeneity of the protein samples he had examined, nor for the accuracy of the analytical methods he had employed. Although I expressed my doubts to Niemann, an action that markedly diminished the warmth of our friendship, I do not recall having spoken to Bergmann about my skepticism, for in those years our relationship was rather formal. It thawed greatly after 1938, when I had shown that pepsin can catalyze the hydrolysis of synthetic peptides, and by that time Bergmann had abandoned the periodicity hypothesis. As in the case of the Svedberg theory, immediate criticism of the Bergmann-Niemann hypothesis was muted, and it was not until 1942 that it was expressed in print by Chibnall.[75] Niemann left the Bergmann

73. Bergmann and Zervas (1932).
74. Fruton (1982a).
75. Chibnall (1942, 1966). See also Witkowski (1985).

laboratory in 1937, but in 1939 there appeared a paper by Pauling and Niemann with the following passage:

> Considerable evidence has been accumulated suggesting strongly that the stoichiometry of the polypeptide framework of protein molecules can be interpreted in terms of a simple basic principle. This principle states that the number of each individual amino acid residue and the total number of all the amino acid residues contained in a protein molecule can be expressed as the powers of the integers two and three.[76]

This paper was largely devoted to a criticism of the "cyclol" hypothesis of protein structure, which received much attention during the 1930s and received the support of the Nobel Prize winner Irving Langmuir:

> We may sum up the present position with regard to the structure of proteins as follows. A vast amount of data relating to protein structure have been collected by workers in a dozen different fields. No reasonable doubt remains as to the chemical composition of proteins. The original idea of native proteins as long chain polymers of amino-acid residues, while consistent with the facts relating to the chemical composition of proteins in general, was not a necessary deduction from these facts. Moreover, it is incompatible with the facts of protein crystallography, both classical and modern, with the phenomena of denaturation, with Svedberg's results which show that the native proteins have definite molecular weights, and with the high specificity of proteins discovered in studies in immunochemistry and enzyme chemistry. All these facts seem to demand a highly organized structure for the native proteins, and the assumption that the residues function as two-armed units leading to long-chain structures must be discarded . . . The agreement between the properties of the globular proteins and the cyclol structures proposed for them is so striking that it gives an adequate justification for the cyclol theory, especially in view of the

76. Pauling and Niemann (1939), p. 1867.

fact that this great variety of independent facts are on this theory seen to be logical consequences of one simple postulate.[77]

The cyclol hypothesis was advanced by Dorothy Wrinch, a British mathematician who had been a member of the "Biotheoretical Gathering" and who had become interested in the structure of proteins, then considered to be the most important biological macromolecules. Astbury had previously inferred from his X-ray diffraction data for fibrous proteins (keratin, collagen) the existence of hexagonal folds, and one of the possible covalent structures had been suggested by the physicist Frederick Charles Frank. Wrinch developed these ideas into a hypothesis, based largely on topological considerations, that a native globular protein is a cage-like structure constructed of six-membered rings formed by the covalent union of the NH group of one amino unit with the CO group of another amino acid unit to produce =N-C(OH)= bonds. During the 1920s, the idea that hexagonal units (other than those present in the amino acids phenylalanine, tyrosine, and tryptophan) are significant structural elements of proteins guided the research of some organic chemists, including Emil Abderhalden and Max Bergmann, who were influenced by Emil Fischer's ambiguous commitment to the linear peptide theory of protein structure in his suggestion that diketopiperazines are possible protein constituents. In these respects, therefore, it would seem that Wrinch's hypothesis was in accord with a tradition based in part on the authority of Emil Fischer. Moreover, Wrinch drew upon the numerological hypotheses of Svedberg and of Bergmann and Niemann to justify her geometric hypothesis that a unit of 288 amino acids is a significant feature of protein structure. The Wrinch hypothesis is now one of the discarded efforts to explain, through geometry, the complexity of protein structure, but in the recent (and justifiable) assertion of the equality of the human sexes there have been some exaggerations of Dorothy Wrinch's role in the development of knowledge in this field.[78] For example, in my opinion, it will simply not do to ascribe the shortcomings of her hypothesis to "artifacts" in the claims of Svedberg and Bergmann. As one who conversed with her during the 1930s about her views, I soon came to realize, with much sympathy, that her campaign to

77. Langmuir (1939), p. 611; see also Wrinch (1938).
78. Abir-Am (1987a, 1987b).

win acceptance of her hypothesis was an attempt to obtain a secure academic post. My skepticism about her theory of protein structure, as well as about those then expounded by Svedberg and Bergmann, stemmed from the impression that wishful thinking ("preconception of what the truth might be") had prevailed over informed knowledge and critical judgment about the empirical data available at that time concerning the chemical constitution of proteins.

The Wrinch hypothesis of protein structure is only one relatively recent example of the centuries-old search for simplicity and unity in the use of geometry to explain the complexity of natural objects and phenomena. Thus, a straight line was deemed a simpler kind of curve than a circle, which in turn was preferred over an ellipse; and cubes or dodecahedrons appeared in ancient philosophical writings as idealized three-dimensional descriptions of materials found in nature. In more recent times, before the techniques of X-ray crystallography had been refined to a stage where they could indicate the irregular shapes of protein molecules, inferences were drawn from the rates of diffusion of proteins in solution, and the results were expressed in terms of "disymmetry constants" which denoted the deviation of the shape of a protein from a perfect sphere, with idealized pictorial representations of "ellipsoids of revolution." Later, as is well known to all present-day students of biochemistry, the spiral ("helical") structure of polypeptide and polynucleotide chains became a prominent feature of the representation of the possible three-dimensional shape of proteins and of nucleic acids.[79] The importance of geometric models in the development of the conceptual framework of modern organic chemistry cannot be overestimated. Perhaps the best-known example is August Kekulé's depiction, as a hexagon, of his theory of the structure of benzene. It may now be difficult for students of chemistry to understand why some distinguished organic chemists of Kekulé's time, notably Hermann Kolbe, denounced Kekulé's approach to the structure of organic compounds or, for that matter, the van't Hoff–Le Bel stereochemical hypothesis based on Kekulé's idea of a tetravalent carbon atom. Whatever may have been written about the origins and later fate of these hypotheses, there can be no question, I believe, of the historical fact

79. The decisive contributions in this development were presented in the papers by Pauling et al. (1951). For a fanciful "spiral" theory of protein structure, see Krafft (1938). See also Luisi and Thomas (1990).

that they guided the fruitful efforts of experimental organic chemists, from Adolf Baeyer and Emil Fischer to this day.[80]

Clearly, the search for simplicity through numerical or geometric representations of the complexity of natural objects and processes has been an essential part of the methodology of the chemical and biochemical sciences. It may be asked, however, to what extent the value of such idealized representations of hypotheses should be judged in terms of their logical coherence, as compared with other criteria. In general, such a representation has been an approximate apprehension of "what the truth might be," rather than the "solution" of a chemical or biochemical problem, and the most valuable models have been those which raised new challenges to the participants in a particular area of research. More often than not, such challenges have been met by the invention of new experimental methods and instruments and by the insight and skill of the practitioners in that area. If in the formulation of a hypothesis about a complex problem, such as that of the structure of proteins, account is not taken of the limits of the methods and instruments which provided the relevant empirical data, the exercise of logical thought and mathematical skill may be fruitless in stimulating further productive experimental effort. As chemists have learned to their disappointment, inherited ideas about the "regularity" of the structure of a particular class of substances do not always stand up to further critical experimental test. For example, in one of the first reports on the three-dimensional structure of the protein myoglobin, it was stated that "there appears to be a real simplicity of chain structure in myoglobin, which will perhaps be shown by other favourably built proteins, and which makes it particularly suitable for intensive X-ray investigation."[81] Eight years later, after such "intensive X-ray investigation," Kendrew and his associates reported that "perhaps the more remarkable features of the molecule are its complexity and its lack of symmetry. The arrangement seems to be almost totally lacking in the kind of regularities which one instinctively anticipates, and it is more complicated than has been predicted by any theory of

80. The long list of relatively recent books and articles on the origins of the theory of chemical structure, with emphasis on the contributions of Kekulé, Couper, and Butlerov, includes those by Bykov (1962), Russell (1971), Rocke (1981), and Wotiz and Rudofsky (1987); see also Gerratt (1987).

81. Bragg et al. (1950), p. 356. For a brief account of the beginnings of protein crystallography, see Perutz (1985).

protein structure."[82] In recent years, protein crystallographers have frequently referred to a model derived from the analysis (and refinement) of X-ray diffraction data as the "solution" of the structure of a protein. Apart from the possibility that the protein under study may undergo changes in its three-dimensional structure when it is not part of a crystalline array of molecules, the meaning of the word "solution" is obviously different from that in mathematical usage.

To the uncertainties faced by the lone investigator who seeks simplicity in scientific explanation, especially when preconceived ideas or anticipations of the outcome of experiments may influence the manner in which experiments are conducted, must be added the perils which attend the work performed by a closely directed group of students and assistants led by a senior person who generates the ideas upon which the research effort is based. Whether it be out of a sense of loyalty, or ambition for academic preferment, junior associates may be tempted to "cook" experiments to produce the result predicted by the leader of the group. In the recent past there has been much brouhaha about "fraud in science," and committees have drafted pious documents about scientific ethics.[83] The number of identified cases of such misbehavior in the biochemical sciences is undoubtedly smaller than those which have actually happened, and the probability of their occurrence has increased with the growth of the scientific population and with the proliferation and size of "biomedical" research groups. It is not a new phenomenon, however, for earlier in this century Emil Abderhalden and Fritz Kögl published reports of work in which wishful thinking appears to have played a large part.[84] Indeed, some of the most imaginative modern scientists have recognized the danger inherent in scattering their hypotheses among a group of loyal students and research assistants. As an example, I cite the answer van't Hoff gave when his student von Euler complained that he had not been assigned a more challenging research problem. According to von Euler's later account, "He [van't Hoff] maintained that a group of young scientists would possibly follow him in hypothetical paths on account of his authority and his former successes, and that he would have a heavy responsibility as an academic teacher concerning the scientific edu-

82. Kendrew et al. (1958), p. 665.
83. See Ayala et al. (1989), Mazur (1989), and Chargaff (1986), pp. 193–205.
84. Karlson (1986).

cation of a new generation."[85] I commend this statement to the attention of my present-day colleagues in the biochemical sciences.

Possibly as a consequence of the recognition, by some contemporary philosophers, of the irrelevance to scientific practice of their professional discourse about the "logic of scientific discovery" or the "logic of justification," closer consideration has been given to the value of "detailed and articulate attention to experiment."[86] One of the case histories, offered by Ian Hacking and deemed to reveal seemingly new features of experimental research, was drawn from the biochemical sciences. This study dealt with the efforts of Roger Guillemin and Andrew Schally to determine the structure of a chemical constituent of the mammalian hypothalamus (thyrotropin releasing hormone, TRH) which was known to stimulate the secretion of a pituitary hormone (thyrotropin, thyroid stimulating hormone, TSH) which, in turn, stimulates the secretion of the thyroid hormone thyroxin.[87] I will not relate the details of the outstanding work performed by the groups led by Guillemin and Schally, except to note that the sequence of their research followed a course of investigation which had become, by the 1960s, the preferred procedure for the determination of the constitution of a physiologically active chemical substance. At each stage of the procedure, many difficulties had to be overcome, especially in the development of a reliable assay for the test of the physiological activity of TRH during the course of attempts to isolate it from its biological source and in the preparation of a sufficiently large purified sample to allow inferences to be drawn about its chemical nature and composition. In the case of TRH, the amounts which were isolated were too small to permit the kind of investigation conducted by Sanger in his studies on insulin, even with the new analytical methods then available. As it happened, the chemical examination of TRH suggested that it is a relatively simple tripeptide, composed of units of glutamic acid, histidine, and proline. Since the number of possible isomers (or their derivatives) was limited, the Guillemin and Schally research groups undertook to synthesize by known chemical methods (largely devised by peptide chemists during the 1950s and 1960s) as many of

85. Euler (1952), p. 644.
86. Hacking (1988), p. 293.
87. Latour and Woolgar (1986); see also Wade (1981).

such compounds as possible. These massive experimental efforts led to the finding of one substance which proved to possess a physiological potency close to that of natural TRH. In my view, however, Hacking's discussion of the methodological implications of this important episode is an example of the neglect by a philosopher of science of the continuity of the historical development of the chemical and biochemical sciences and, in this case, of seeming indifference to the background of the development of endocrinological research and the art of peptide synthesis. Although I must add that recent issues of philosophical journals contain case histories in which fuller account is taken of the history of research on other biochemical topics, such as the vitamin theory,[88] it would seem that some present-day professional philosophers have not yet accepted the view that a profitable discussion of a recent achievement in the biochemical sciences depends on a thorough familiarity with its historical background and with the present state of development of the interplay of theory and experiment in the area of inquiry under consideration.

Such familiarity has been evident in many twentieth-century philosophical writings about the methodology in areas of scientific inquiry (for example, planetary mechanics, optics, electromagnetism, equilibrium thermodynamics) which had ceased to be active fields of research effort. As Hanson noted, however, "In a growing research field, inquiry is directed not to rearranging old facts and explanations into more elegant formal patterns, but rather to the discovery of new patterns of explanation."[89] In the context of the importance of new observation and experiment, and the invention of new methods and instruments for such discovery, it is difficult to disagree with Feyerabend's statement that "scientists do not solve problems because they possess a magic wand—methodology, or a theory of rationality—but because they have studied a problem for a long time."[90] I welcome, therefore, the efforts of some professional philosophers of science to counter, through the informed and detailed study of modern scientific practice, the aristocratic intellectual attitude which places imagination and reason above observation and experimentation.[91]

It cannot be doubted that during the decades before 1900, when

88. Akeroyd (1988).
89. Hanson (1958), p. 2.
90. Feyerabend (1975), p. 302.
91. See Franklin (1986), Galison (1987), Gooding et al. (1989), Golinski (1990), and Baird and Faust (1990).

"natural philosophy" and "natural history" were being transformed into separate research disciplines, serious attention was paid by many prominent scientific workers to the views and precepts of professional philosophers regarding the "scientific method." During the twentieth century, however, there appears to have been an increasing alienation between analytical philosophy and scientific practice, and I believe that this recent trend deserves closer examination by future historians of Western culture.[92]

Obviously, one of the reasons for the indifference of most of my scientific colleagues to the writings of present-day philosophers is that the present social climate of research in the biochemical sciences, with its intense competition in fashionable fields, hardly encourages a philosophical attitude among the people actively at work in these fields. A more important reason, in my opinion, is the enormous acceleration during this century of the growth of reliable knowledge in diverse areas of scientific inquiry with various methodologies whose character has been changing as a result of the interplay of theory and practice. To those practitioners in growing disciplines, such as the biochemical sciences, who have taken note of the efforts of some leading philosophers of science to formulate logical principles of reasoning which are independent of time or of the nature of the problems under investigation, these efforts do not appear to offer much promise outside the realm of continued discourse among professional philosophers.

In this realm, the two dominant twentieth-century groups have been the Anglo-American linguistic philosophers and the "Vienna Circle" of logical positivists. The writings of the first group are distinguished by their elegance in the use of "plain english" in the logical analysis of complex philosophical concepts, but they have largely avoided the knotty problems of the methodology of scientific research.[93] For example, in declining to judge the disputes about the relationship of the physical and biological sciences, Gilbert Ryle stated that

> I have no first-hand and very little second-hand knowledge of the specialized ideas between which these systems of thought are braced. I have long since learned to doubt the native sagac-

92. This trend has been discussed by Gale (1984), among others.

93. Perhaps the most notable example is the book by Ayer (1952). See also the collective volume of essays edited by Lewis (1963).

ity of philosophers when discussing technicalities which they have not learned to handle on the job, as in earlier days I learned to doubt the judgements of those towing-path critics who had never done any rowing.[94]

In contrast, the logical positivists were more ambitious in their philosophical aspirations. One of the disciples of the Vienna school was Karl Popper, who, after seceding from the school and emigrating to England, became the most famous British philosopher of science. In a review of Popper's book *Conjectures and Refutations*, Ayer wrote:

In the world of scholarship, as in politics, the sharpest conflicts tend to be fratricidal. So Popper has always been at pains to emphasize his disagreement with the logical positivists, whose general standpoint might strike a detached observer as having much in common with his own. He objects to their picking on the possibility of verification, rather than the possibility of falsification, as the standard of empirical content; he objects to their treatment of their principle of verification as a criterion of meaning; he objects to the attempt which some some of them have made to secure a basis for knowledge in sensory experience, and indeed to any attempt to justify claims to knowledge in terms of their sources: above all he reproaches them for being inductivists. For he holds not only that inductive reasoning can not be logically justified, but that the scientific method does not require it. . . . A welcome admission which Popper does make in this essay is that it is required of a good theory not merely that it should be subject to new tests, but that it should sometimes succeed in passing them. . . . But now, since Popper also holds that its success in passing such tests corroborates a given hypothesis, one may well ask how his position differs from that of the inductivists who regard the consonance of a hypothesis with new evidence as confirming it.[95]

I have quoted this passage to indicate the views of a noted British philosopher regarding Popper's ideas about the "scientific method"

94. Ryle (1954), p. 12. A stronger statement is that "the Darwinian theory has no more to do with philosophy than has any other hypothesis of natural science"; Wittgenstein (1922), pp. 76–77.

95. Ayer (1963), p. 155; see Popper (1962).

and the relation of these ideas to those advanced by the logical positivists of the Vienna Circle. To a skeptical biochemist, both the linguistic philosophy ("talk about talk") and the Popperian variety ("argument about argument") appear to fall short of providing useful guides to understanding the interplay of thought and practice in the chemical and biochemical sciences. I find hope, however, in the writings of present-day philosophers who have taken heed of the seminal approaches of Hanson and Kuhn.[96] In particular, I hope that they will expand the horizon of their inquiry to include the informed and critical study of the historical development of these sciences.

To sum up my views in this chapter, I believe that the growth of knowledge about natural objects and phenomena is a centuries-old, complex, uneven, and heterogeneous process, characterized by the continuous interplay of thought and practice. Among the forms of thought have been logical reason and non-logical imagination, and among the forms of practice have been the invention, use, and modification of instruments, as well as skill and craftsmanship. Because what we now call scientific research has emerged from the uneven development of the various areas of this human endeavor, some sectors of scientific inquiry appear to be more "advanced" than others. Because of the complexity and heterogeneity of the problems presented by natural objects and phenomena, the development of the methodology of scientific inquiry has undergone change at different rates in the various areas of scientific research. Because scientific research is a social activity, all aspects of the growth of scientific knowledge—the choice of problems, the recognition of discovery, the evaluation of facts and theories—have been influenced by the nature of the environment in which such research was conducted. Although I recognize that some kinds of deductive logic are essential tools of scientific thought, I find in some kinds of inductivism much that describes actual practice in the chemical and biochemical sciences. In short, I doubt whether there is any such thing as the "scientific method," independent of time and circumstance, but rather tend to believe that there are many such methodologies, all of which have emerged from the interplay of thought and action within separate areas of inquiry, and which have continued to change in response to challenges presented by new problems, or the restatement

96. In addition to the books mentioned in note 5 of Chapter 1, I call attention to the papers of Giere (1989) and Nickles (1987).

of old problems, within those individual areas. Some philosophers of science may find these opinions somewhat naive, but I wish to assure them that I do not subscribe to Pascal's dictum, "se moquer de la philosophie, c'est vraiment philosopher." Rather, I subscribe to the view of Bertrand Russell:

> A philosophy which is to have any value should be built upon a wide and firm foundation of knowledge that is not specifically philosophical. Such knowledge is the soil from which the tree of philosophy derives its vigour. . . . Philosophy cannot be fruitful if divorced from empirical science. And by this I do not mean only that the philosopher should 'get up' some science as a holiday task. I mean something much more intimate: that his imagination should be impregnated with the scientific outlook and that he should feel that science has presented us with a new world, new concepts and new methods, not known in earlier times, but proved by experience fruitful where older concepts and methods proved barren.[97]

To this counsel, I add that of the philosopher Jonathan Cohen, who, in writing of inductive judgment, stated:

> Any inductive judgement is itself empirically corrigible. Conclusive certainty is never inductively justifiable. But that is a deprivation that we can live with. To keep our minds open to the possibility of new evidence is all that it requires of us, and while we do this we are still entitled to claim an appropriate degree of justification for the theories that have survived the most thorough tests which we at present believe ourselves capable of devising. Human rationality requires us at any one time to do the best we then can, not to do the best that could ever be possible.[98]

97. Russell (1959), pp. 230, 254.
98. Cohen (1989), p. 187.

CHAPTER THREE

The Interplay of Biology and Chemistry

I begin this chapter with a quotation which I used to conclude my book *Molecules and Life,* written before the advent of the present age of biotechnology. The last section of that book dealt with some aspects of the history of the centuries-old problem of "the whole and the parts." The quotation was taken from the published version of a lecture presented in 1922 by the noted biologist Edmund Beecher Wilson, who, after summarizing the new knowledge provided by cytologists, geneticists, and chemists about the parts of the living cell, asked:

> How shall we put it together again? It is here that we first fairly face the real problem of the physical basis of life; and here lies the unsolved riddle. We try to disguise our ignorance concerning this problem with learned phrases. We are forever conjuring with the word "organization" as a name for the integrating and unifying principle in the vital processes; but which one of us is really able to translate this word into intelligible language? We say pedantically—and no doubt correctly—that the orderly operation of the cell results from a dynamic equilibrium in a polyphasic colloidal system. In our mechanistic treatment of the problem we commonly assume this operation to be somehow traceable to an orginal pattern or configuration of material particles in the system, as is the case with a machine. Most certainly conceptions of this type have given us an indispensable working method—it is the method which almost alone is

> responsible for the progress of modern biology—but the plain fact remains that there are still some of the most striking phenomena of life of which it has thus far failed to give us the most elementary understanding. . . . We are ready with the time-honored replies: It is the "organism as a whole"; it is a "property of the system as such"; it is "organization." These words, like those of Goldsmith's country parson, are of "learned depth and thundering sound." Once more, in the plain speech of everyday life, their meaning is: *We do not know*. . . Shall we then join hands with the neo-vitalists in inferring the unifying and regulatory principle to the operation of an unknown power, a directive force, an archaeus, an entelechy or a soul? Yes, if we are ready to abandon the problem and have done with it once for all. No, a thousand times, if we hope really to advance our understanding of the living organism. To say *ignoramus* does not mean that we must also say *ignorabimus*.[1]

Wilson's role in the development of experimental biology in the United States cannot be overestimated.[2] Apart from his distinguished personal researches and those of his students (notably Walter Sutton) in cytology, embryology, and genetics, Wilson was a profound and critical observer of these areas of biological investigation. He was also an outstanding teacher, as is clear from the first edition of his book *The Cell in Development and Inheritance*, in which he strove "to bring the cell-theory and the evolution-theory into organic connection."[3]

The third edition (1925) of Wilson's book appeared at a time when biologists who sought chemical explanations for the "physical basis of life" were more strongly influenced by the concepts of physical chemistry, as expounded by Wilhelm Ostwald, and by those of his followers who developed colloid chemistry, than by the organic chemistry of their time. In large part, this preference may be attributed to the fact that the proteins, long considered by biologists to be the essential components of "protoplasm," appeared to be studied more successfully by colloid chemists than by those who followed

1. Wilson (1923), pp. 284–286.
2. See Muller (1943), Baxter (1974), and Allen (1976).
3. Wilson (1896), p. 11.

Emil Fischer's organic-chemical approach to the problem of protein structure.[4]

Wilson's book reflected the state of empirical knowledge in cytology, embryology, and genetics that had grown out of the outstanding achievements in these fields in the preceding decades. That period was also characterized by debates, involving not only biologists but also philosophers and theologians, on the question whether the phenomena exhibited by living things can be "explained" in terms of the properties of the chemical substances which had been isolated from the tissues and fluids of biological organisms, and in terms of the physical forces which were then known to determine the structure and interaction of these substances. The debate continued after 1925, but gained new intensity after the transformation of biochemical thought and practice when experimental support was available for the Watson-Crick model of DNA as an explanation of the chemical mechanism for the transmission of hereditary characters. The latter part of this chapter will deal with some of the recent writings on this problem. Since these writings, for the most part, have tended to ignore the historical continuity of the debate, I believe it is helpful to recall important previous episodes in the interplay of biological and chemical thought and practice which elicited controversy about many of the same issues in dispute today.

The Nineteenth-Century Debates

In many ways, the recent excitement appears to resemble that which occurred a century earlier in Victorian England, when Thomas Henry Huxley led a campaign to establish "biology" as a unified discipline based on the idea that the phenomena exhibited by living things are the consequence of the properties of their cellular protoplasm, whose nature is entirely chemical and which obeys the known laws of physics.[5] In addition, several German biologists, notably Ernst Haeckel and Carl Nägeli, offered similar "mechanistic" speculations about the nature, organization, and development of the

4. The several editions of Wilson's book merit more attention than they appear to have received from recent writers on the "philosophy of biology" or the "logic of life." Notable exceptions include the book by Mayr (1982), pp. 647, 812, 817, and the article by Roll-Hansen (1978).

5. Huxley (1869). See Geison (1969) and Caron (1988).

matter of which living things were then thought to be composed. In the foreground of the debates during the last four decades of the nineteenth century were questions relating to the cell-theory, as developed during the period 1840–1870, and the theory of biological evolution, as expounded by Darwin and Wallace in 1858–1859. Huxley and Haeckel, each in his own way, espoused and defended Darwin's theory, but it is, I believe, fair to say that Haeckel's ideas about the implications of the two theories were more influential in guiding subsequent biological research, especially in the study of embryonic development and of heredity.[6]

The cell-theory, in the form presented by Matthias Schleiden and Theodor Schwann, was a statement of a hypothesis about the fundamental unity of the diverse forms of living things, and an attempt to define the differences between "inorganic" and "organic" matter. In addition to asserting that animals and plants are composed of similar elementary units, with nuclei and cell walls, Schwann speculated about the manner in which the formed elements arise from a structureless fluid which he called *cytoblastema*.[7] In this speculation he reflected the ancient philosophical tradition, attributed to Thales, that solid bodies arise from fluids, as well as more recent research such as that of Henri Dutrochet, who, during the 1820s, had attempted to produce contractile fibers by the treatment of solutions of albumin or gelatin with agents such as electricity, acid and alkali, or by desiccation.[8] Also, the idea of a structureless fluid had appeared in the writings (1835) of the protozoologist Felix Dujardin, who named it *sarcode*. Schwann suggested, in a lengthy section of his famous book, that the formation of cells is analogous to the growth of inorganic crystals, and Schleiden later wrote:

> Even if one is not prepared to accept Schwann's analogy between cell and crystal and considers this as for the time being entirely unfounded, there is nevertheless available in this clever

6. Haeckel (1866). For accounts of Haeckel's life and work, see Heider et al. (1919), Uschmann (1961), and Heberer (1968). Haeckel's influence is evident in later work in cytology (notably that of Richard Hertwig), embryology (Hans Driesch), and especially in marine biology (for example, Anton Dohrn).

7. Schwann (1839). For the early development of the cell-theory, see Baker (1948–1955) and Klein (1936).

8. For Dutrochet, see Schiller and Schiller (1975), Pickstone (1976), and Schiller (1980), pp. 131–156.

> exposition the undeniable possibility, that one day science will successfully explain the cell as an equally necessary form, in a relatively solid state, of a permeable (assimilated, organic) substance transformed according to law, just as crystals are transformed from impermeable (inorganic) substances. All organisms which are formed or reproduced as individual single cells would then represent particular species of organic crystallization.[9]

The cell-theory of Schleiden and Schwann was extensively modified by later biologists, especially after the demonstration that cells arise from preexisting cells, but the crystal analogy retained its attractiveness for those who sought chemical models of the organization and development of living things. Thus, during the latter half of the nineteenth century much attention was paid to the "micellar" theory of the botanist Carl Nägeli. This theory, first presented during the 1860s in a description of starch grains as crystalline arrays of molecules, was later used by Nägeli to formulate a distinction between the *trophoplasm* (the nutritive matter of egg cells) and the *idioplasm*, which bears the hereditary elements of sperm cells.[10] Moreover, during the 1920s and 1930s, some biologists became greatly interested in "liquid crystals," such as those formed when some organic compounds (for example, cholesteryl benzoate) are heated to their initial melting point. This phenomenon had been studied intensively by the physicist Otto Lehmann, who found that such liquid crystals retained a three-dimensional structure similar to that of solid crystals.[11] Thus, in 1936 Joseph Needham stated:

> The aspect of molecular pattern which seems to have been most grossly underestimated in the consideration of biological phe-

9. Schleiden (1850), vol. 2, p. 519; see Lorch (1974). Schwann's crystallization hypothesis was questioned repeatedly after Jakob Henle and Karl Reichert had rejected it during the 1840s. For valuable accounts of the development of crystallography before the 1830s, see Metzger (1918), Burke (1966), Goodman (1969), Mauskopf (1976), and Burckhardt (1988).

10. Nägeli (1884). See Wilkie (1960–1961) and Robinson (1979), pp. 109–130. For Nägeli, the distinctive property of crystalline structures was their capacity for "intussusception," whereas for Haeckel it had been "imbibition"; see also Haeckel (1917) on the "souls" of crystals.

11. See Lehmann (1907, 1911) and Przibram (1926). The influence of these writings on several embryologists has been discussed by Haraway (1976).

> nomena is that found in liquid crystals. . . . Living systems actually *are* liquid crystals or, it would be more accurate to say, the paracrystalline state undoubtedly exists in living cells.[12]

Another example is the proposal by Erwin Schrödinger in 1943 that the gene be thought of as an "aperiodic crystal."[13] The tenacity with which the crystal analogy has been retained in biological thought is suggested by François Jacob's acceptance of the idea that although it may not be appropriate to compare the organization of the chemical components of biological systems to the growth of a three-dimensional inorganic crystal, all that is needed is to redefine the term "crystal" to include repetitive two-dimensional structures, such as membranes, or even one-dimensional ones such as fibers:

> If the constraints for the formation of three-dimensional crystals seem to be strict, they appear to be less rigid in the other cases. When the sub-units of nucleic acids or proteins represent sufficiently identical objects [they] order themselves into geometrical arrangements. An entire series of biological structures, the polymers, the membranes, the organelles distributed in the cell, thus have their natural logic.[14]

It may be questioned, however, whether the fact that some of these repetitive structures yield X-ray diffraction data from which features of regularity may be inferred justifies the revival of the crystal analogy as applied to biological systems.

Although crystalline structures were observed during the middle of the nineteenth century in plant cells, and the red coloring matter of mammalian blood had been crystallized, at the time of Huxley's famous address (and for decades afterward), more attention was paid to the biological importance of the non-crystalline "colloidal state of matter," as Thomas Graham termed it. Graham described the colloidal state as "a dynamical state of matter, the crystalloidal being the static condition. The colloid possesses ENERGIA. It may be

12. Needham (1936), pp. 156–157.

13. Schrödinger (1944); see Yoxen (1979). Schrödinger's purported role in the emergence of molecular biology is discussed in the next chapter.

14. Jacob (1970), p. 325. This translation differs from the one in the English version; see Jacob (1973), p. 304.

looked upon as the probable primary source of the force appearing in the phenomena of vitality."[15]

By the 1860s, the jelly-like *sarcode* of Dujardin had become a *protoplasm* present in the cells of both plants and animals, and composed of colloidal "albuminoid" substances which, in Huxley's words, formed the "physical basis of life."[16] Huxley's views were widely publicized, attacked, and defended, and the ensuing debate raised the intensity of nineteenth-century controversy among persons whom later historians have dubbed "vitalists," "materialists," or "mechanists." That controversy took various forms, depending on the particular area of biological inquiry in which it was conducted. One arena concerned the morphology and physiology of organisms which were only visible with the aid of an optical microscope of the time. After the widespread acceptance of the cell-theory, especially in the the form enunciated by Max Schultze during the 1860s, the question of the status of such microscopic bodies in the hierarchy of the biological world received animated attention. Are they, as Christian Gottfried Ehrenberg had claimed during the 1830s, complex living things, with a full complement of physiological systems comparable to those in higher organisms? Or are they primitive forms of the cells of plants and animals, whose principal morphological feature is the presence of an albuminoid protoplasm?[17] The great cytological achievements during the latter years of the nineteenth century in the description of the changes in cell nuclei associated with the union of ova and spermatozoa, and in the nuclear chromosomes during cell division, while providing an empirical groundwork for the study of the mechanisms which might be operative in the transmission of hereditary characters from parents to progeny in the plant and animal kingdoms, left open many questions about the biology of the microorganisms. Indeed, in the interplay of chemical and biological thought and experiment during the nineteenth century, some of the most intense controversies arose from the demonstration in 1837, by Schwann and by Charles Cagniard-Latour, that the fermentation of glucose to alcohol by yeast is associated with the activity of living organisms.[18]

15. Graham (1861), p. 184.
16. See Güttler (1972).
17. The course of this debate has been discussed by Churchill et al. (1989) in a series of papers on the history of protozoology.
18. See Fruton (1972), pp. 42–86.

In the formulation of his cell-theory, Schwann distinguished between the "plastic" processes of growth, which he likened to crystallization, and the "metabolic" processes in which food materials are converted to cell substance and excretory products. Before his publications on alcoholic fermentation and the cell-theory, Schwann had described in 1836 the agent present in gastric juice which converts dietary proteins into more readily soluble forms. He named this agent, discovered two years earlier by Johann Nepomuk Eberle, *pepsin*.[19] During the 1830s similar agents, in extracts of germinating barley and in human saliva, were found to liquefy starch paste with the formation of soluble sugars. The names *diastase* and *ptyalin* were assigned to these agents, and in succeeding decades *lipase* (which converts fats to glycerol and fatty acids) or *maltase* and *invertase* (which cleave disaccharides to their component sugars) were added to the list of what came to be called "soluble ferments." At mid-century such agents were taken to be chemical "catalysts" (a term introduced by Berzelius in 1838) which effect the hydrolytic cleavage of the principal items of the animal diet—proteins, carbohydrates, and fats. What was in dispute was the question whether analogous catalysts are present in yeast cells, and are responsible for the fermentation of sugars. Moreover, the processes in animal organisms in what was variously called metabolism, metamorphosis, nutrition, or *Stoffwechsel* were considered to be of two distinct kinds. One kind, about which there was some experimental knowledge, was collected under the common heading of disassimilation (later catabolism), and included chemical processes in which nutrients and cellular constituents are broken down by hydrolysis, fermentation, or oxidation. The other kind of metabolic processes, termed assimilation (later anabolism), was thought to involve the condensation of small molecules to more complex ones, or the reversal of the intracellular processes associated with fermentation or oxidation. Thus, in addition to the question whether the fermentation of glucose by yeast involves chemical agents comparable to pepsin (in 1878 Kühne dubbed them "enzymes"), or whether such fermentations, as Louis Pasteur maintained, are intimately linked to the life of microorganisms, there was the larger question whether the anabolic (in later terms, biosynthetic) phases of metabolism are also exclusively biological, rather then chemical, aspects of the

19. Schwann (1836); see Hickel (1975).

physiology of living things. During the nineteenth century, the first of these questions occasionally aroused bitter controversy among leading chemists and biologists, largely on the ground that no one had succeeded in obtaining a cell-free soluble material that could effect the conversion of glucose to alcohol. This dispute abated, but was not wholly stilled, after Eduard Buchner, an organic chemist, reported in 1897 that he had isolated, from yeast press-juice, a protein-like substance which fermented glucose and which he named *zymase*.[20]

To one of the other questions, about the mechanisms whereby living organisms effected the biosynthesis of complex cell constituents, the answer of Claude Bernard was most incisive. After his important discoveries of the release of glucose from the liver (1848) and the formation of glycogen (a polysaccharide) in the liver (1857), Bernard wrote:

> The first entirely vital action, so termed because it is not effected outside the influence of life, consists in the creation of the glucogenic material in the living hepatic tissue. The second entirely chemical action, which can be effected outside the influence of life, consists in the transformation of the glycogenic material into sugar by means of a ferment.[21]

Later, Bernard generalized this conclusion by stating that in living organisms there are

> . . . two kinds of phenomena, one of organization or assimilation, the other of disorganization or disassimilation. . . . Both kinds of phenomena are equally physiological, they are equally necessary for the maintenance of life. Assimilation, which is in reality nothing but a sort of reduction, cannot occur without disassimilation, which is in reality a combustion. . . . We have in the liver the two kinds of phenomena of assimilation and of disassimilation. The phenomena of assimilation correspond to the formation of glycogen, the phenomena of disassimilation correspond to its transformation into dextrin and glucose. One

20. Kohler (1971, 1972).
21. Bernard (1857), p. 583; see Fruton (1972), pp. 405–411.

of these phenomena is no more post-mortem than the other; they both occur in the living organism.[22]

Bernard developed this theme more fully in his book on the phenomena of life common to animals and plants,[23] and elsewhere he also noted that

> it is a great mistake to believe that the physico-chemical phenomena in the organism are the same as those which occur outside it, and to have wished to explain these phenomena by agents which have been identified outside it. It is into this error [that] some chemists, who reasoned from the [chemical] laboratory to the organism, have fallen, whereas it is necessary to reason from the organism to the laboratory. . . . From what we have said above that every chemical phenomenon which occurs in the organism cannot be effected by the same inorganic agent, but by another organic agent created by life, it follows that the same chemical phenomenon has two causes: one inorganic, the other organic. . . . Therefore, I suppose that [Marcelin] Berthelot makes genuine fat by synthesis and with strong inorganic reagents, [but] this could not at all prove that things happen this way in the organism.[24]

Thus, as part of his effort to establish animal physiology as a discipline independent of the clinical medicine of his time, Bernard also repeatedly criticized the physiological speculations of organic chemists such as Liebig, Dumas, and Berthelot. Moreover, because Bernard considered the fermentation of glucose to be a degradative process, he declined to accept Pasteur's dictum that such fermentations are correlative with the life of microbial cells.[25]

During the latter half of the nineteenth century, the debates about the role of enzymes in the dynamics of intracellular processes occasionally became heated, as in the exchange between Felix Hoppe-Seyler and Willy Kühne, or in the assertion of the physiologist Eduard Pflüger, who espoused a theory of "energy-rich" protoplasmic proteins, that the assumption of intracellular enzymes was

22. Bernard (1877), pp. 360–362.
23. Bernard (1879), vol. 2, pp. 211ff.
24. Bernard (1947), pp. 242–243.
25. See Pasteur (1879).

"not only unnecessary but indeed highly implausible."[26] In particular, even after Buchner's preparation of zymase in 1897, and after some oxidative enzymes had been obtained in soluble form, physiologists repeatedly called attention to the fact that no enzyme had been shown to effect the synthesis of a cellular constituent.[27] This attitude was widespread among biologists until the 1930s, and indeed vestiges of it appeared during the 1950s with the hypothesis (from Jacques Monod's laboratory) that in living cells proteins are only synthesized, and that protein breakdown occurs only after cellular death.[28] For the Belgian biochemist Hubert Chantrenne, writing in 1966, "Biochemistry in 1938 looked somewhat depressing to a student eager to learn about the processes of life." After summarizing some of the high points of the subsequent developments in the study of biosynthesis, especially through the contributions of Fritz Lipmann, Chantrenne noted that "biochemistry is no longer the chemistry of death and decay, it is the chemistry of the living cell, with its essentially irreversible, oriented processes, admirably organized and controlled."[29]

The quotations given above from the writings of Claude Bernard suggest that in the nineteenth-century controversy about "vitalism" versus "materialism" or "mechanism," he was (as Owsei Temkin has termed him), a "vitalistic materialist" and quite different in his philosophical outlook from the "mechanical materialists" of the German group that included Hermann Helmholtz, Emil du Bois-Reymond, Ernst Brücke, Adolf Fick, and Carl Ludwig.[30] Although all these men appreciated, as did Claude Bernard, the value of the available chemical knowledge in the study of physiological processes in animal organisms, the scientific questions they asked and their experimental approach were rather different from those of Bernard. Whereas Bernard considered experimental surgery and the administration of toxic agents, such as carbon monoxide or curare, to be the most valuable methods in physiological studies, the Ger-

26. For the controversy between Hoppe-Seyler and Kühne, see Fruton (1990), pp. 82–92. For Pflüger's statement, see Pflüger (1878), p. 249; a similar view was expressed by Nägeli (1879), pp. 86–87.

27. See, for example, Neumeister (1903), pp. 78–79.

28. Hogness et al. (1955); see also Mandelstam (1958).

29. Chantrenne (1966), p. 37.

30. Temkin (1946); Schiller (1967), pp. 201–227. See also Cranefield (1957), Benton (1974), Galaty (1974), and Rudolph (1983).

man group sought to apply and develop the theoretical and experimental physics of their time for this purpose. During the middle years of the nineteenth century the achievements of the members of this group outshone, in many respects, those of their contemporaries who were seeking explanations of physiological phenomena in the language of the organic chemistry of their time. The physical theories of mechanics, heat, optics, acoustics, and electricity served as the basis for physiological experiment, and new instruments were invented which greatly broadened the scope of biological investigation. For a time, this approach came to be called "biophysics," and important contributions were made to the understanding of such physiological processes as muscular movement, transmission of nerve impulses, and vision, as well as the dynamics of respiration, secretion, and the circulation of the blood.[31] In addition to the many lasting theoretical and experimental contributions of the German "mechanistic" physiologists of the nineteenth century, the influence they exerted in other countries was particularly evident in Britain and the United States. Thus, there was a remarkable succession of brilliant investigators (from Charles Scott Sherrington and John Newport Langley to Edgar Douglas Adrian, Joseph Erlanger, and Herbert Spencer Gasser) in neurophysiology and in the physiology of muscle (from Archibald Vivian Hill to Andrew Fielding Huxley and Hugh Esmor Huxley). It is noteworthy that, although in their own research most of these physiologists of the nineteenth and early twentieth centuries relied largely on the use of physical instruments, they also recognized the importance of the work of chemically-minded biologists, and encouraged their efforts. Both Carl Ludwig and Emil du Bois-Reymond had, in their institutes, independent sections of physiological chemistry, as did the pathologist Rudolf Virchow. Later, in England, Sherrington and Walter Morley Fletcher were influential in the establishment of independent university departments of biochemistry.

The Emergence of Biochemistry

Among the men who came from such biological environments were Felix Hoppe-Seyler and Franz Hofmeister in Germany and Frederick

31. For historical accounts of the development of animal physiology, see Brooks and Cranefield (1959) and Rothschuh (1973). After the Second World War, the term "biophysics" was adopted by physicists who sought to engage in biological research; for a comment about this trend from a noted physiologist, see Hill (1956).

Gowland Hopkins in England. They, in turn, bred a generation of investigators who sought to explain physiological phenomena in terms of the organic chemistry of their time, and in some instances (for example, Eugen Baumann) these biochemists made valuable contributions in that field. During the period 1900–1930, their efforts began to make evident the complexity of the chemical nature of protoplasm, and of the metabolic processes in living organisms. They demonstrated the inadequacy of such hypothetical entities as the "energy-rich" protoplasmic protein of Eduard Pflüger or the "living" aldehyde protein of Oscar Loew and the "biogen" of Max Verworn, and, as proposed by Frederick Gowland Hopkins in 1913, recognized that

> in the study of the intermediate processes of metabolism we have to deal, not with complex substances which elude ordinary chemical methods, but with simple substances undergoing comprehensible reactions. By simple substances I mean such as are of easily ascertainable structure and of a molecular weight within a range to which the organic chemist is well accustomed.[32]

By the early 1930s that range had been extended, and through the interaction of organic chemists and biochemists significant advances had been made in the study of the constitution of proteins and nucleic acids; new biochemical entities such as "co-enzymes," hormones, and vitamins had been identified; and the problems of the "intermediate processes" of metabolism were attacked by investigators such as Franz Knoop, Gustav Embden, and Otto Meyerhof. What had been only plausible surmises about the pathways in the metabolic breakdown and synthesis of such relatively simple substances such as glucose, amino acids, or urea were replaced by conclusions drawn from ingenious experimentation in which the methodology of organic chemistry was combined with those of various biological disciplines.[33] I mention here only two of the numerous examples which might be cited: one of them was the work (1907) of Walter Fletcher and Hopkins on the production of lactic acid during the anaerobic contraction of frog muscle and the disappear-

32. Hopkins (1913), p. 214. For theories about "living" proteins, see Pflüger (1875), Loew (1896), and Verworn (1903).
33. See Fruton (1972), pp. 397–485.

ance of lactic acid in the presence of oxygen during the period of recovery;[34] the other came in 1932, in the work of Hans Krebs (with Kurt Henseleit), on the biosynthesis of urea in slices of rat liver.[35] As I have noted elsewhere, for the first time a biochemical synthesis was explained in terms of chemical reactions identified in the appropriate biological system, and not merely inferred by analogy to the known chemical behavior of the presumed reactants.[36]

These and other achievements during the first decades of this century, such as those of Heinrich Wieland, Thorsten Thunberg, David Keilin, and Otto Warburg in the study of biological oxidations, marked out new approaches and raised new problems for experimental study. New techniques markedly furthered the effort to examine in greater detail the chemical pathways in the dynamics of living organisms. The one which transformed the methodology of biochemistry most radically, and which today still represents an essential feature of experimentation in the biochemical sciences, was the use of compounds labeled with isotopes of the principal chemical elements of biological systems. Before the Second World War, Rudolf Schoenheimer and his associates had demonstrated the power of the isotope technique in metabolic experiments through the use of compounds labeled with the stable isotopes deuterium (^{2}H) and ^{15}N, provided by the chemist Harold Clayton Urey. After the war, when radioactive isotopes became avaliable, the labels of choice became tritium (^{3}H), ^{14}C, ^{32}P, and ^{35}S. Also, during the 1940s, George Beadle and Edward Tatum introduced the use of mutant strains of the mold *Neurospora crassa* into the study of metabolic pathways, and later other organisms (especially *Escherichia coli*) were extensively employed for this purpose. The vast extent of the biochemical knowledge provided by means of these two techniques cannot be summarized here.[37] What must be stressed, however, is that apart from their value in supporting earlier hypotheses about several metabolic pathways, and in disproving others, their use revealed the occurrence of previously unexpected biochemical reactions and invited the closer study of the intracellular apparatus

34. Fletcher and Hopkins (1907).

35. Krebs and Henseleit (1932); Krebs (1981), pp. 51–60. An important account of the discovery of the "ornithine cycle" of urea synthesis has been prepared by Holmes (1991).

36. See Fruton (1972), p. 436.

37. See Florkin (1979) and Fruton (1972), pp. 445–485.

involved in these reactions, particularly through the isolation and characterization of the individual enzymes which catalyze these reactions. In this respect, the more recent concentration on the intracellular processes associated with the replication of DNA, the DNA-directed synthesis of RNA (and its reversal), the RNA-directed synthesis of proteins, and other chemical reactions (restriction, ligation) studied by molecular biologists has reflected the biochemical tradition inherited from work during the 1930s of Northrop and Warburg, and their associates, on the isolation and purification of enzymes.[38] As I indicated in the previous chapter, the introduction of new or improved techniques for the separation of biochemical substances, both large and small, was decisive in the advances made after the Second World War, and the use of X-ray diffraction analysis, nuclear magnetic resonance spectroscopy, and other physical methods made possible the determination of the chemical structure and the study of the interactions of such substances.

Moreover, it should be noted that during the period 1900–1930 biochemical problems had once again became prominent in the efforts of organic chemists. During the middle years of the nineteenth century, the area of scientific inquiry which Berzelius had termed "organic chemistry," because he considered it to deal with the substances and processes in the tissues and fluids of biological organisms, had largely severed its connection with biology. In part, this separation was a consequence of the rejection by leading physiologists of the biological speculations, based principally on analytical data for the elementary composition of foodstuffs and excretory products, offered during the 1840s by Justus Liebig and Jean Baptiste Dumas. A much more important reason for the separation was the change in the conceptual framework of organic chemistry brought about by the acceptance of new ideas about the mode of linkage (valence) and spatial arrangement (stereochemistry) of the atoms which constitute organic compounds. The fruitfulness of these theories was evident in the rapid development during the second half of the nineteenth century of many new methods for the synthesis, in the chemical laboratory, of both "natural" and "unnatural" organic substances, including dyes and drugs. In Germany, the ties of university chemists to the burgeoning chemical industry played a major role in this development. Some leading organic chemists ap-

38. See Kornberg (1989).

pear to have deplored the growing separation of organic chemistry from biology. In 1859 Emil Erlenmeyer wrote: "If a man like Ludwig identifies the progress of physiology with the progress in chemistry, and finds in chemistry, because of its ability to explain the most delicate and complex processes, part of the salvation of physiology, then this must be a stimulus toward physiological investigations."[39] Although Adolf Baeyer early in his career showed an interest in biochemical problems, and his most brilliant pupil, Emil Fischer, undertook the study of sugars and purines and later the proteins, it was not until after about 1910 that other leading organic chemists began intensive biochemical research.[40] The most prominent ones were two other former members of the Baeyer school: Heinrich Wieland and Richard Willstätter. Later in the twentieth century, the role of organic chemists in the elucidation of the structure of biochemical substances such as porphyrins, steroids, vitamins, and antibiotics was decisive in the study of intracellular processes, and they also made important contributions to the investigation of the mechanisms of enzymatic catalysis.

Nineteenth-Century Cytology, Embryology, and Microbiology

During most of the nineteenth century, the stimulus for the interplay of biology and chemistry had come largely from the growth of knowledge about such physiological processes as respiration, the metabolic conversion of food materials, vision, muscular contraction, or microbial fermentation. The scope of the interplay was broadened after about 1870 by significant new developments in biological thought, observation, and experimentation, especially in cytology, embryology, and microbiology. These developments were spurred by important technical advances in the design of microscopes and microtomes, and in the use, as cytological stains, of natural dyes such as carmine or hematoxylin and of synthetic dyes such as eosin or methylene blue.[41] In particular, the staining technique revealed remarkable changes in components of cell nuclei during the course of the fertilization of eggs by spermatozoa, cell

39. Erlenmeyer and Schöffer (1859), p. 316.
40. Fruton (1990), pp. 154–158, 235–241.
41. For a history of microscopy, see Bradbury (1967). For the development of cytological staining techniques, see Clark and Kasten (1983).

division (mitosis), and the maturation of the germ cells (meiosis). The nuclear components which bound the basic stains were termed "basophilic." Thus, during the 1870s, Oscar Hertwig and Hermann Fol demonstrated that, in the sea urchin *Toxopneustes lividus*, fertilization involved the union of the nuclei of the two germ cells. Also, Walther Flemming showed that the basophilic strands (later to be called "chromosomes") in cell nuclei undergo striking changes during mitosis, and called attention to the fact that the stainable material, which he named *chromatin*, had been found by Eduard Zacharias to resemble the *nuclein* previously isolated by Friedrich Miescher. Flemming's observations were confirmed and extended during the 1880s by Eduard van Beneden and Theodor Boveri, who used the worm *Ascaris megalocephala*, and by Eduard Strasburger for nuclei of plant cells. These investigators described in greater detail the linear structure and apparent longitudinal division of the chromosomes during cell division. In addition, it had been shown that the number of chromosomes is constant in a species and that, in meiosis, the chromosomes are composed of corresponding members derived from both parents.[42] By the end of the 1880s many biologists had accepted the view of Wilhelm Roux and August Weismann that the chromosomes of the parent germ cells (gametes) are responsible for the transmission of hereditary characters, and shortly afterward Weismann's prediction that, in meiosis, the chromosome number is halved ("reduction division") was confirmed in cytological studies. By that time, it was also known from the chemical work of Albrecht Kossel that Miescher's nuclein is composed of a basophilic "nucleic acid" (a term introduced by Richard Altmann in 1890) and a protein component, to form a "nucleoprotein." As Wilson expressed it in 1895, "inheritance may perhaps be effected by the physical transmission of a particular compound [nucleic acid] from parent to offspring."[43]

For the discoverer of nuclein, however, the evidence offered by the "guild of dyers" (as Miescher termed them) was unacceptable. Although he did not publish his opinions, they were expressed in letters to his uncle, Wilhelm His, and appeared posthumously in a

42. For a history of cytology, see Hughes (1959); see also Coleman (1965).

43. Wilson (1895), p. 4. However, in the third edition of his famous book he rejected this opinion and called attention to the uncertainties in the staining methods used to follow the fate of chromatin; see Wilson (1925), p. 351.

volume edited by His and published in 1897.[44] His, then professor of anatomy at Leipzig, had earlier espoused the idea of "transmitted motion" in fertilization: "It is neither the form, nor form-building material that is transmitted, but the excitation to form-developing growth."[45] In his famous paper of 1874 on the nuclein from the sperm of the Rhine salmon, Miescher had echoed this view:

> There are no specific fertilization substances. The chemical phenomena have only a secondary significance; they are subordinate to a higher explanation. If we seek an analogy to explain all the available knowledge, it seems to me that there remains nothing but a picture of an apparatus that evokes or transforms some kind of motion.[46]

In letters to His, written in 1892 and 1893, Miescher offered a stereochemical theory of heredity based on the presence of "numerous asymmetric carbon atoms in organized substances" and dismissed Weismann's views as "afflicted with half-chemical concepts which are partly unclear, and partly correspond to an obsolete state of chemistry."[47] If Miescher's earlier view reflects the "physicalist" effort of his uncle to bring the approach of the "mechanistic" physiologists into embryology, his later chemical speculations do not appear to be any clearer than those of the "guild of dyers" whom he criticized, especially in his seeming lack of recognition of the contributions of Kossel to the elucidation of the chemical nature of the nuclear constituent which Miescher had discovered.

At the seventy-second meeting of the German society of scientists and physicians, held at Aachen in 1900, Oscar Hertwig summarized the development of biology in the nineteenth century by claiming that the greatest triumphs were achieved in his own field of microscopic anatomy, through the discovery with the aid of the compound microscope that plants and animals are constructed of "innumerable tiniest elementary organisms," and that from the cell-theory and the protoplasm-theory biology has derived a firm foundation, just as chemistry is based on the atomic-molecular theory. According to Hertwig,

44. Miescher (1897).
45. His (1874), p. 152.
46. Miescher (1897), vol. 2, p. 98.
47. Ibid., vol. 1, pp. 117, 122.

> More thorough research, paired with philosophical insight, teaches more clearly year by year that the cell, this elementary keystone of living nature, is far from being a peculiar chemical giant molecule or even a living protein and as such is likely eventually to fall prey to the field of an advanced chemistry. The cell is itself an organism, constituted of many smaller units of life.[48]

Although Hertwig paid due homage to Darwin's theory of natural selection, this affirmation of the autonomy of biology on the basis of the success of cytology may be compared with more recent claims, to be discussed later in this chapter, based on the success of the neo-Darwinian "evolutionary synthesis."

During the four decades which separated the publication of Charles Darwin's *On the Origin of Species* and the recognition ("rediscovery") of the importance of Gregor Mendel's work during the 1860s on the hybridization of plants, cytological and biochemical observations had provided some clues to the possible processes whereby the morphological and physiological properties of biological species are transmitted from parents to offspring. These processes were widely considered to be linked to the mechanisms involved in the transformation of a fertilized egg into a highly differentiated multicellular organism. Although Haeckel's famous dictum, usually abbreviated in the statement that "ontogeny recapitulates phylogeny," enjoyed great popularity and stimulated fruitful efforts in comparative descriptive embryology, by the end of the century it was the experimental study of developmental physiology which attracted greater attention. A leading figure was Wilhelm Roux, who found in 1888 that after he had punctured the nuclei of fertilized frog eggs, cellular differentiation continued with the production of "half-embryos." Three years later, Hans Driesch succeeded in separating (by shaking) the first two cells ("blastomeres") formed from a fertilized sea urchin egg, and showed that each cell could develop into a complete, but small, embryo at the gastrula stage. Near the end of the century, Jacques Loeb found that unfertilized sea urchin eggs, after exposure to relatively high concentrations of sodium chloride and return to ordinary sea water, could undergo "artificial parthenogenesis" up to the larval stage. These

48. The quotations are from Querner (1972), p. 187.

embryological experiments, along with others (notably those of Curt Herbst), appeared to support Roux's view that the area of biological inquiry which he named *Entwicklungsmechanik* (developmental mechanics) should be explored on the basis of the assumption that the organization and regulation evident in embryological development arise solely from the physico-chemical constitution of biological organisms. This mechanistic attitude was opposed during the 1890s (and afterward) by Driesch, who shifted from *Entwicklungsmechanik* (and science altogether) to vitalism and enunciated his teleological concept of "entelechy." At the turn of the century, therefore, the problems posed by the growing knowledge of the complexity of the physiological processes in the transformation of a fertilized egg into an adult organism made the field of developmental biology one of the principal arenas of a controversy which has continued into recent times.[49]

If, before 1900, the problems presented by the cell-theory and the evolution-theory provided the principal challenges to biologists, those raised by the closer study of unicellular organisms also generated much dispute. The arguments in which Louis Pasteur played a prominent role, such as those about fermentation or "spontaneous generation" or the microbial cause of infectious disease, are well known, as are those in which Robert Koch was involved. The French adulation of Pasteur was matched by that accorded in Germany to Koch for his role in demonstrating the microbial cause of anthrax, tuberculosis, and cholera. There can be no doubt that for the subsequent development of bacteriology, the contributions of Pasteur and Koch to medical practice were decisive, and stimulated the establishment of the research institutes bearing their names. Because of their public prominence, however, Pasteur and Koch have overshadowed other major nineteenth-century figures in the emergence of what came to be called "microbiology."[50] Among these men was

49. For a recent treatment of the history of embryology, see Horder et al. (1986); apart from essays on several aspects of the development of this field, the volume includes a valuable bibliography of earlier published sources. Among these, the volumes by Needham (1959) and Oppenheimer (1967) are especially useful. For a detailed account of Driesch's views, see Churchill (1969).

50. For the history of microbiology, see Bulloch (1938) and Collard (1976). The list of writings about Pasteur and Koch is enormous; for admirable biographical articles and lengthy lists of other sources, see Geison (1974) and Dolman (1973).

Ferdinand Cohn, who began as a botanist with an interest in algae and fungi. He then attempted to bring order into the classification of the bacteria, and his discovery of bacterial spores markedly affected the course of the debate about spontaneous generation.[51] Two others were Martinus Henricus Beijerinck and Sergei Nikolaevich Winogradsky, who developed the "enrichment" method for the isolation of pure cultures of known bacteria and for the discovery of new bacterial forms. Their studies on the physiology of bacteria which can utilize inorganic nitrogen and sulfur compounds for growth laid the groundwork for a vast expansion of biochemical knowledge, beyond that gained from the investigation of the metabolism of higher animals and plants. Moreover, in 1898 Beijerinck concluded from his research on the mosaic disease of tobacco plants that the pathogen (described by Dmitri Ivanovski in 1892) is a noncellular self-reproducing infectious fluid *(contagium vivum fluidum)*, because it could pass through a filter which retained bacteria. In the discussion of his findings on the tobacco mosaic virus, Beijerinck wrote:

> Although the reproduction or growth of a dissolved particle is not unthinkable, it is difficult to imagine. Molecules equipped with a division mechanism enabling them to reproduce, and the idea of metabolizing molecules which must be a presupposition, seems to me obscure, if not positively unnatural. Hence it might conceivably serve as an explanation that the contagium, in order to reproduce, must be incorporated into the living protoplasm of the cell, into whose reproduction it is, in a manner of speaking, passively drawn. This would at least reduce two riddles to only one, since the incorporation of a virus into the living protoplasm, even if well-documented, cannot by any means be considered a thoroughly understandable process.[52]

51. For a biography of Cohn, see Geison (1971).

52. Beijerinck (1899), p. 31; the English translation is that given by Wilkinson (1976), pp. 116–117. For Beijerinck, see Iterson et al. (1940); for Winogradsky, see Waksman (1953). A valuable discussion of the role of microbiology in the development of biology was prepared by Kluyver and van Niel (1956).

Twentieth-Century Embryology versus Genetics

I have restricted most of the foregoing summary to the development of biological knowledge, and its interplay with chemistry, during the nineteenth century in order to emphasize the importance for some areas of controversy of Mendel's studies on the hybridization of the garden pea. As has been recounted in different ways by innumerable scientists, historians, and philosophers, the "rediscovery" in 1900 of Mendel's findings by the botanists Carl Correns and Hugo de Vries marked the beginning of significant shifts in the direction of biological research. Several aspects of these changes are, I believe, of interest in relation to the interplay of biology and chemistry during the twentieth century.

One prominent feature of the "rediscovery" of Mendel was the disparity between what he had done and written and the interpretation of his findings in light of the cytological knowledge gained after 1870, as well as of the debates about theories (such as those of Darwin and Haeckel) which linked the mechanisms of inheritance and embryonic development to biological evolution.[53] What emerged from that interpretation was a chromosomal theory of inheritance based on the transmission of elementary units (akin to Mendel's *Elemente,* Weismann's *Biophore,* or de Vries's *pangens*). During the 1860s, however, in the absence of the knowledge gained after he had ceased to do scientific work, Mendel's objective does not appear to have been to study the mechanisms of inheritance, but to formulate and test a relatively simple theory which could predict the quantitative outcome of the crossbreeding of varieties of the same plant species. Apart from his botanical knowledge and skill, Mendel had been educated in the physics and chemistry of his time, and no doubt had come to appreciate the importance of mathematics in the formulation of scientific theories. As has been suggested by several biographers of Mendel, it seems likely that he had developed the outlines of his celebrated hypotheses of the "segregation" and "independent assortment" (as they were later named) of hereditary characters before selecting *Pisum sativum* and partic-

53. The many accounts of Mendel's work and its "rediscovery" include those of Roberts (1929), Stern and Sherwood (1966), Olby (1966), Orel (1984), and Callender (1988). A comparison of such accounts reveals many differences in emphasis and interpretation; see, for example, those of the geneticist Dunn (1965), pp. 3–24, and of the evolutionary biologist Mayr (1982), pp. 710–726.

ular pairs of alternative ("dominant" or "recessive") traits as his test system, after preliminary trials had indicated that experiments with this plant and these "alleles" might yield numerical data in accord with these hypotheses. It also seems probable that he refined his hypotheses during the course of the trials, and that he and his assistant were biased in favor of these hypotheses in judging the validity of individual observations, such as whether a particular seed was round or wrinkled. In short, Mendel's methodology does not appear to have been very different from the practice of productive chemists and physicists. By attempting to avoid complexity and striving for simplicity, Mendel gave to the biologists of 1900 a set of principles which appeared to offer a means of reducing the complexity of the phenomena observed during the preceding three decades. The fact that Mendel himself was disappointed to find from experiments (suggested by Nägeli) with the hawkweed *Hieracium* that inheritance in this plant did not conform to his hypotheses and that, after 1900, many such "exceptions" were found and explained in terms of a more complex version of the chromosome theory, only confirms the view that the search for simplicity in the form and function of biological organisms has been most fruitful when it has led to the discovery of new complexity.

The Mendelian-chromosome theory was not greeted with universal acclaim by leading biologists, largely because that theory assigned to the cytoplasm, a term introduced during the 1850s by Albert Kölliker to denote the non-nuclear cellular contents, a secondary role in the processes whereby the morphological and physiological traits of the parents appear in their offspring during the course of embryonic development. The roots of the ensuing controversy lie deep in antiquity, but after the enunciation of the cell-theory the issue was stated clearly in 1861 by Ernst Brücke, a member of the group of physiologists later identified as "mechanistic materialists":

> As advisable as it may be to depend constantly and strictly on what is immediately observable, so it is necessary that the intellectual eye not be closed to what is inaccessible to observation, in order that we do not overrate our microscopic perception and, with the help of such catchwords as cell membrane, cell content and cell nucleus construct physiological doctrines to which a later generation may deny recognition. . . . We cannot

> think of any living vegetating cell with a homogeneous nucleus and a homogeneous membrane and with a plain albuminoid solution as [its] content, because we do not detect in the properties of albumin, as such, the phenomena we denote as the phenomena of life. We must therefore ascribe to living cells, apart from the molecular structure of the organic compounds which they contain, still another structure of different complexity, and this we denote by the name of organization.[54]

In succeeding decades, the use of improved microscopic and staining techniques revealed that the cytoplasm is not an undifferentiated "albuminoid solution" but that it contains structurally distinct units such as those later named mitchondria and microsomes. This recognition of the complexity of what had been termed "protoplasm" and new observations of the process of embryonic development led some leading embryologists to reject the view that this process is solely controlled by the nuclear chromosomes. For example, Edwin Grant Conklin wrote:

> From all sides the evidence has accumulated that the chromosomes are the principal seat of the inheritance material; until now this theory practically amounts to a demonstration. On the other hand, all persons who have studied cell-lineage have been impressed with the fact that polarity, symmetry, differentiation and localization are first visible in the cytoplasm and that the positions and properties of the embryonic parts are dependent upon the location and sizes of certain blastomeres or cytoplasmic areas.[55]

Some years later, Jacques Loeb expressed the view that "the facts of experimental biology strongly indicate the possibility that the cytoplasm of the egg is the future embryo (in the rough) and that the Mendelian factors only impress the individual (and variety) characters upon the rough block."[56] In keeping with the biochemical knowledge of his time, Loeb suggested that the Mendelian factors

54. Brücke (1861), pp. 385–386.
55. Conklin (1905), p. 220.
56. Loeb (1916), p. 8.

"can impress only individual characteristics, probably by giving rise to special hormones and enzymes."[57]

If Loeb's "mechanistic" outlook on biological problems led him to speculate along biochemical lines, other embryologists sought explanations for differentiation, regulation, and organization by invoking the existence of "metabolic gradient fields." This theory was proposed by Charles Manning Child during the 1920s, adopted by several British biologists, and later more fully developed by Paul Weiss.[58] Apart from these and many other speculations about embryonic development, after 1900 there were many important microscopic observations, such as those of Santiago Ramon y Cajal and Hans Held on the histological changes in the development of the nervous system, and significant experimental studies, most notably those of Ross Granville Harrison and of Hans Spemann.[59]

In 1907 Harrison introduced into embryological research the technique of aseptic tissue culture, by means of which he was able to answer some of the questions raised by earlier histological observations about the dynamics of the formation of nerve fibers. This method was later adopted by investigators (among them Alexis Carrel and Albert Fischer) in other areas of experimental biology.[60] Moreover, Harrison had been among the first to begin the use and elaboration of the method of transplantation of small portions of an embryo ("heteroplastic grafting") discovered by Gustav Born in 1897. Harrison and his students used this method, principally with embryos of the salamander *Amblystoma punctulata*, to study the development of various structures such as the limb, eye, or ear, and to seek explanations for the stereospecificity of the formation of such bilateral structures. Although he sought physico-chemical explanations in the writings of Lehmann and Przibram on liquid crystals, and even (with William Astbury) in the use of X-ray diffraction analysis, Harrison eschewed the kind of extreme physicalist speculation prevalent in the embryology of his time.[61] In my opinion, Harrison represents an outstanding example of a great scientist who

57. Ibid., p. vi.

58. Weiss (1947).

59. For recent discussions of the researches of Harrison and Spemann, see Horder et al. (1986), pp. 149–177 [by J. A. Witkowski], and pp. 183–242 [by T. J. Horder and P. J. Weindling], respectively. For Harrison, see also Maienschein (1983).

60. See Witkowski (1979, 1980a).

61. Oppenheimer (1972); Haraway (1976), pp. 64–100.

explored, with skill and acumen, one of the central problems of biology with the best experimental methods he could use or devise, but who, in offering speculations during the 1930s about the mechanisms of embryonic development based on the "configuration of the protein molecule and its accompanying chemical and physical activities,"[62] may not have appreciated fully the limits of the knowledge available to him about the structure and "activities" of proteins. In this respect, Harrison's views reflect the continuity of an effort by "mechanistic" biologists to seek in the chemical opinions of their time explanations of the complexity of biological processes. The choices they made were frequently rather uncritical. Among other examples which might be cited, I will only note that the distinguished founder of developmental genetics, Richard Goldschmidt, welcomed the Bergmann-Niemann hypothesis of protein structure as an important contribution to the theory of the gene, and that, at an earlier time, when the status of enzyme chemistry was uncertain and controversial, Hans Driesch (before he became a vitalist) proposed that the cell nucleus discharges *Fermente* into the cytoplasm.[63]

Hans Spemann, the other principal figure in the development of experimental embryology during the first decades of this century, became famous (he received the 1935 Nobel Prize in Physiology or Medicine) largely through his enunciation of the biological principle of an "organizer" which causes "induction" in the differentiation of embryonic cells. In 1921, experiments by his student Hilde Pröscholdt (later Mangold) showed that cells from the dorsal lip of an amphibial blastophore, after transplantation into the ventral region of a second gastrula, developed into neural tissue.[64] During the course of the ensuing experimental work by Spemann's associates (notably Johannes Holtfreter) and others, it was found that, after being heated, the ventral ectoderm itself acquired the capacity to act as an organizer, thus suggesting the existence of a chemical material which might be extracted and identified. Among those who sought during the 1930s to isolate the organizer, and for a time thought it to be a steroid (like other known hormones), were Joseph Needham and Conrad Hal Waddington. Another biochemical pos-

62. Harrison (1937), p. 372.

63. Goldschmidt (1938); Driesch (1894), pp. 88–91.

64. Horder et al. (1986), pp. 261–284 [by L. Saxen and S. Toivonen]; Hamburger (1984, 1988).

sibility appeared to be an effect of the organizer in changing the rate of intracellular oxidative processes, for Needham and Waddington had found that methylene blue (a non-biological oxidation-reduction indicator used during the 1930s in studies of cellular respiration) also can cause "induction." By the end of the decade, however, it was clear that this biochemical approach was not likely to contribute significantly to the understanding of the mechanisms of embryonic development. Nor did the lively discusions among members of the "Biotheoretical Gathering," which included Needham, Waddington, Joseph Henry Woodger, John Desmond Bernal, and Dorothy Wrinch, lead to fruitful experimental efforts in the field of embryology. Some of the members of this group drew particular inspiration from the topological speculations of D'Arcy Wentworth Thompson about biological growth and form.

At the end of the 1930s, therefore, it seemed to many biologists that the efflorescence of the kind of experimental embryology initiated by Wilhelm Roux's *Entwicklungsmechanik* had reached an impasse. Although there was no shortage of theoretical speculations, especially in Britain where the philosophical bias toward "organicism" received impetus from the writings of Alfred North Whitehead, some younger embryologists felt the need for experimental approaches which, in addition to those used by Harrison and Spemann, also included the ideas and methods of the genetics of their time. This attitude reflected a reaction to the separation, after about 1910, of the study of the transmission of hereditary characters from that of embryonic development, as a consequence of the continuing controversy over the primacy of the nucleus over the cytoplasm, or vice versa. For Harrison, writing in 1937, "The prestige of success enjoyed by the gene theory might easily become a hindrance to the understanding of development by diverting our attention solely to the genome, whereas cell movements, differentiation, and, in fact, all developmental processes are actually effected by the cytoplasm."[65]

So much has been written about the success of the gene theory that only a brief reminder of some aspects of its history after 1900 is necessary here. The first cytological evidence for the applicability of Mendel's principles of segregation and assortment to the chromosomal theory of inheritance was provided independently in

65. Harrison (1937), p. 379.

1902–1903 by Theodor Boveri and by Walter Sutton, when the latter was working in the laboratory of Edmund Wilson.[66] This evidence was not widely accepted, however, and greater attention was paid to the application, by Hugo de Vries, of Mendel's theory to the problem of "mutation" in the evening primrose *(Oenothera)*, which produced many "mutant" varieties. In his extended studies of this phenomenon, de Vries was interested both in the fundamental problem of the mechanisms of biological evolution and also in the application of Mendelian theory to agricultural practice. It was the latter aspect of the field, to which William Bateson gave the name "genetics,"[67] that generated the greater public excitement, especially among plant and animal breeders in the United States, where the achievements of Luther Burbank were widely admired. American agriculturists came to believe that the application of Mendelian theory might lead to the improvement of the quality and quantity of plants and animals developed through hybridization. There began, by 1910, a flow of financial support from the U.S. Department of Agriculture, state legislatures, and private agencies (notably the Carnegie Institution of Washington) to university departments of plant breeding and to agricultural experiment stations which brought, over the years, a succession of outstanding investigators into the field of genetics. Among those who made fundamental contributions in that field were such later notables as Lewis John Stadler, George Beadle, and Barbara McClintock. Nor was the investment without economic profit, for it produced valuable varieties of hybrid corn and wheat.[68]

De Vries's studies on *Oenothera* also appear to have stimulated the embryologist Thomas Hunt Morgan to search for mutations in the vinegar fly, *Drosophila melanogaster*, an organism introduced into genetics during the period 1901–1906 by William Ernest Castle. Until 1910, Morgan had rejected the Mendelian theory, but in that year "in a pedigree culture of Drosophila which had been running for nearly a year through a considerable number of generations, a male appeared with white eyes. The normal flies have red eyes."[69]

66. For Boveri, see Baltzer (1962); for Sutton, see McKusick (1960).

67. Bateson (1909). In declining to accept the evidence for the chromosomal theory of inheritance, Bateson offered instead a modernized version of the physicalist theory of His, mentioned earlier.

68. Rosenberg (1976), pp. 196–209. The rise of genetics also spawned the growth of an organized "eugenics" movement in the United States, as well as in other countries; see Ludmerer (1972) and Glass (1986).

69. Morgan (1910), p. 120.

From hybridization studies, Morgan concluded that the "factor" for red eyes is closely associated with the "factor" for maleness, and that his data accorded with Mendel's theory. It is noteworthy that, in his 1910 report, he did not use the word "gene," introduced in 1909 by Wilhelm Johannsen, nor for that matter the word "chromosome." There ensued, however, a rapid change in Morgan's attitude toward the new genetics based on cytological observation and breeding experiments, the conversion no doubt abetted by his friend and departmental chief Edmund Wilson, who had been among the first to welcome the "rediscovery" of Mendel. As has been recounted many times, the appearance of the white-eyed mutant marked the beginning of a memorable series of studies in Morgan's "fly room" at Columbia University by a small research group which included Alfred Sturtevant, Calvin Bridges, and Hermann Muller.[70] After his initial discovery, Morgan studied double mutants of *Drosophila* and found evidence for gene recombination, which he explained as a consequence of the "crossing-over" of homologous chromosomes. These results led to a more precise theory of a linear order of genes in each chromosome, and to Sturtevant's famous experiments in which the relative position of the genes was estimated by determining the frequency of crossing-over for pairs of sex-linked factors. Also, in a masterly combination of breeding experiments and cytological examination, Bridges explained an apparent exception by demonstrating the "non-disjunction" of the X-chromosomes (which carry the sex-linked factors) in the female *Drosophila* at meiosis. Later, such interplay between cytological and breeding studies provided explanations of many other instances, in a variety of organisms, where the Mendelian principles appeared to be violated. By 1915, the Morgan group had accepted the view that chromosomes are the bearers of hereditary factors because they "furnish exactly the kind of mechanism that the Mendelian laws call for."[71]

The Emergence of Biochemical Genetics

But what of the chemical nature of the chromosomal constituents which participate in this mechanism? Is it the material which Flemming had named "chromatin" and which, at the end of the nine-

70. Sturtevant (1959); Dunn (1965), pp. 139–157; Carlson (1966); Roll-Hansen (1978).
71. Morgan et al. (1915), p. ix.

teenth century, appeared to be identical to the nucleic acid portion of the material Miescher had named "nuclein"? Opinion on this question shifted toward the negative after 1900, and it is significant (and in my opinion admirable) that Morgan, in his final extended discussion of the theory of the gene, avoided the issue altogether and limited his account to the conclusions drawn from breeding experiments and cytological observations.[72] During the period 1910–1930, however, there was no shortage of biochemical speculation about the nature of genes. At one extreme there was the view expressed by Bateson: "The supposition that particles of chromatin, indistinguishable from each other and indeed almost homogeneous under any known test, can by their material nature confer all the properties of life surpasses the range of even the most convinced materialism."[73] If, in hindsight, present-day molecular biologists or philosophers of biology may find this statement to be an expression of a vitalistic attitude, which it is, the words "homogeneous under any known test" should suggest that Bateson was reflecting the biochemical knowledge and, what is most important, the biochemical methodology of his time. Indeed, the biochemical speculations during 1910–1930 about the nature of genes did not center as much on their relationship to chromatin as on that to enzymes. In particular, the studies of the clinician Archibald Edward Garrod (a colleague of both Hopkins and Bateson) had shown that the human hereditary dysfunction known as alcaptonuria could be attributed to an "inborn error of metabolism" in which the breakdown of the protein amino acid tyrosine is blocked at some stage to lead to the accumulation of an intermediate ("homogentisic acid"), which is converted to a black pigment excreted in the urine.[74] Garrod was able to make this chemical statement because a decade earlier the biochemist Eugen Baumann had established the structure of homogentisic acid and had demonstrated its metabolic origin from dietary tyrosine. Also, several biochemists (notably Gabriel Bertrand) had demonstrated the presence, in animal and plant tissues, of enzymes ("tyrosinases") which catalyze the oxidation of tyrosine to black pigments ("melanins"). One may assume that it was this kind of empirical knowledge which provided the basis for the following two statements by Bateson:

72. Morgan (1926).
73. Bateson (1916), p. 542.
74. Garrod (1902, 1909).

> We may draw from Mendelian observations the conclusion that in at least a large group of cases the heredity of characters consists in the transmission of the power to produce something with properties resembling those of ferments. It is scarcely necessary to emphasise the fact that the ferment itself must not be declared to be the factor or thing transmitted, but rather, the power to produce that ferment, or ferment-like body.[75]

> We must not lose sight of the fact that though the factors operate by the production of enzymes, of bodies on which these enzymes can act, and of intermediary substances necessary to complete the enzyme-action, yet these bodies themselves can scarcely be themselves genetic factors, but consequences of their existence.[76]

Thus, in Bateson's view, and in the language introduced in 1909 by Johannsen and adopted by later biologists, enzymes are elements of the *phenotype* for an individual biological organism, whereas the *genotype* represents the set of hereditary factors transmitted from its parents.[77] Indeed, in 1915 Huia Onslow (at the Biochemical Laboratory in Cambridge) reported that recessive whiteness in rabbits is associated with the absence of tyrosinase, and subsequent studies by Sewall Wright on hair color in mammals led him to suggest that "melanin pigment is formed in the cytoplasm of cells by the secretion of oxidizing enzymes from the nucleus."[78] Some speculations went further, however, in proposing that genes are autocatalysts for their self-propagation. This idea, with variations, was discussed between 1910 and 1930 by Arend Hagedoorn, Leonard Thompson Troland, and Hermann Muller.[79] Since, as Muller stated in 1929, "At present any attempt to tell the chemical composition of the genes is only guess-work,"[80] these speculations were based on the physical chemistry of the time, with repeated reference to the analogy to the growth of crystals.[81]

75. Bateson (1909), p. 268.

76. Bateson (1913), p. 86. The somewhat greater biochemical sophistication of the phrasing of this statement, as compared with the previous one, may be surmised to reflect the influence of the views expressed by Hopkins (see note 32).

77. For a discussion of the genotype-phenotype distinction, see Dunn (1965), pp. 88–97; Churchill (1974); Allen (1979), Sapp (1987), pp. 36–42.

78. Wright (1917), p. 224.

79. Hagedoorn (1911); Troland (1917); Muller (1922). See Ravin (1977).

80. Muller (1929), p. 907.

81. Koltzov (1928); Haldane (1932), pp. 156–157. See also notes 11–14.

Far more fruitful for the further development of genetic theory than these speculations was the continued interplay of cytological observation of chromosomal morphology (and its many "aberrations") and breeding experiments, especially after the discovery by Muller that the treatment of *Drosophila* sperm with X-rays

> induces the occurrence of true "gene mutations" in a high proportion of the treated germ cells. Several hundred mutants have been obtained in a short time and considerably more than a hundred of the mutant genes have been followed through three, four or more generations. They are (nearly all of them, at any rate) stable in their inheritance, and most of them behave in a manner typical of the Mendelian chromosomal mutant genes found in organisms generally.[82]

Subsequent cytological studies by Muller and Theophilus Painter showed extensive changes (breakage, inversion, translocation) in *Drosophila* chromosomes after their exposure to X-rays. Also, Stadler reported that the irradiation of barley seeds with X-rays (or exposure to radium emanation) induced mutations, and that the mutation rate was proportional to the radiation dosage. Soon afterward, other agents (ultraviolet light, chemical substances such as the "nitrogen mustards," urethane, and formaldehyde) were also found to cause mutations in *Drosophila* and other organisms.[83] Muller's discovery led to the entry during the 1930s of some physicists into biology; an often-cited paper based on their efforts at that time is the one by Nikolai Timoféeff-Ressovsky, Karl Zimmer, and Max Delbrück, who offered a target *(Treffer)* theory in which it was assumed that a mutation is induced by a single ionization within a molecular volume with an average size of about 1,000 atoms.[84] However, in a later discussion of the model, Stadler wrote:

> This is an impressive picture, but it has been evident for many years that it has no valid relationship to the experimental data from which it was derived. . . . The basis of the model is the assumption that the statistics of observed mutation are in fact the

82. Muller (1927), p. 84. For an account of versions of the concept of mutation, see Stubbe (1965).
83. Auerbach (1967) and Brink (1967), pp. 67–80 [article by C. Auerbach].
84. Timofeeff-Ressovsky et al. (1935).

> statistics of structural alteration of the molecules that constitute the gene-string. The investigations of specific mutations contradict this assumption and show that the model has no basis in reality.[85]

Stadler thus called attention to the ambiguity in the use of the word "gene" as both a unit within a chromosome which undergoes structural alteration and a unit associated with the appearance of a change in a particular phenotypic character.

Apart from the continued fruitful development of genetics during the 1930s through the interplay of cytological and breeding methods, the use of spectrophotometric and chemical techniques, notably by Torbjörn Caspersson and by Jean Louis Brachet, provided quantitative data for the deoxyribonucleic acid (DNA) content of cell nuclei and demonstrated the localization of most of the intracellular ribonucleic acid (RNA) in the structural elements of the cytoplasm. Also, despite the success of the Mendelian chromosome theory of inheritance, new questions were raised during the 1930s about the relationship of nuclear genes to cytoplasmic development. One of the leaders in keeping alive the view that cytoplasmic factors play a significant role in the transmission of hereditary characters was Tracy Sonneborn, who, as a student of Herbert Spencer Jennings, had been introduced into research on the protozoan ciliate *Paramecium aurelia*. With admirable persistence, Sonneborn succeeded (where his teacher had failed) in developing a procedure for the production of "mating types" of this organism and then embarked on a lifelong and fruitful study of cytoplasmic inheritance by means of the cross-breeding methods of traditional genetics.[86] Another leading figure in the continued debate about the primacy of nuclear genes over cytoplasmic factors was the embryologist Boris Ephrussi, who sought during the 1930s to bring together the genetics and biochemistry of his time.[87] His joint studies with the plant geneticist George Beadle on the inheritance of eye color in *Drosophila* led them to identify two genetic factors, each of which appeared to be involved in a stepwise conversion of the amino acid tryptophan to the red eye pigment. There then followed Beadle's famous inves-

85. Stadler (1954), p. 812.
86. Sonneborn (1938); for a valuable biographical account, see Sapp (1987), pp. 87–122.
87. See Sapp (1987), pp. 123–162.

tigation, with the biochemist Edward Tatum, on the genetic control of metabolic reactions in the mold *Neurospora crassa*. As Beadle described it many years later, they

> . . . hit upon the idea of reversing the procedure we had been using to identify specific genes with particular chemical reactions. We reasoned that, if one primary function of a gene is to control a particular chemical reaction, why not begin with known chemical reactions and then look for the genes that control them? In this way we could stick to our specialty, genetics, and build on the work chemists had already done. The obvious approach was first to find an organism whose chemical reactions were well-known and then induce mutations in it that would block specific identifiable reactions.[88]

In my opinion, few statements in the scientific literature exemplify more clearly the intimate interplay of biological and chemical thought. The biological groundwork for the choice of *Neurospora* had been laid by the previous research of Bernard Dodge and Carl Lindegren, and the chemical groundwork had been provided by the available knowledge of some pathways of intermediary metabolism. Although the relationship between heredity and metabolism had been clear to many (especially British) biochemists since Garrod's writings on alcaptonuria, the work of Beadle, Tatum and their students opened a field of inquiry (biochemical genetics) which owed its success to the choice of microorganisms as the objects of experimental study.

The apparent delay in the emergence of microbial genetics as a major area in the interplay of biology and chemistry may, in retrospect, be attributed in large part to the fact that between 1870 and 1940 more public attention was paid to the medical "microbe hunters" than to biologists and biochemists who studied the growth and metabolism of microbial cultures. After the enunciation of the germ theory of disease by Louis Pasteur and Robert Koch, leading medical bacteriologists considered the principal challenges to lie in such questions as the nature of the "toxins" produced by pathogenic agents and of the "antitoxins" elaborated by animal organisms to

88. Beadle (1963), p. 13.

counteract the effect of these agents. Chemically-minded investigators, such as Paul Ehrlich, sought by hit-or-miss methods to find synthetic "magic bullets" which would control a disease by selectively killing pathogenic organisms. Before about 1940, the few instances in which these efforts proved to be partially successful were outnumbered by the failures, and such "chemotherapy" was not grounded on the knowledge of microbial metabolism that had been accumulating under the stimulus of the work of Beijerinck and Winogradsky and the later research, between the two world wars, of such investigators as Albert Jan Kluyver and Marjory Stephenson. Indeed, it may said that a "rational pharmacology" in the treatment of infectious disease developed only after 1940 through the systematic biochemical study of the action of antibiotics (sulfanilamide, penicillin) which had been discovered either by accident or through the screening of many microbial forms isolated from soil samples.[89]

The emergence, after 1940, of bacterial genetics as a central area of the interplay of biology and chemistry has been recounted in various ways by many writers, and I shall not attempt to summarize what can be found in modern textbooks of biochemistry.[90] Several aspects of the historical background, however, merit recollection. Perhaps the most important one was the growing recognition after 1900 of the metabolic flexibility of microorganisms in response to changes in their chemical environment. Studies such as those of Frédéric Vincent Dienert in 1900 on the fermentation of glucose and galactose by the yeast *Saccharomyces cerevisiae*, or of Henning Karström in 1930 on the utilization by various bacteria of different sugars as the sole organic components of culture media, or of Isaac Lewis in 1934 on lactose metabolism in *Escherichia coli*, called into question the view of the fixity of bacterial species that was implicit in the germ theory of disease as it had been expounded by Louis Pasteur and Robert Koch. Indeed, Dienert's work at the Pasteur Institute provided one of the stimuli for the memorable doctoral research there of Jacques Monod on "diauxic" bacterial growth.[91] The further elaboration of Monod's program led to the discovery of a complex genetic and metabolic apparatus for the induction (or repression) of

89. See Parascandola (1980).
90. See, for example, Stryer (1988); Singer and Berg (1990).
91. Dienert (1900); Monod (1942). See Lewis (1934) and Amsterdamska (1987).

the synthesis of bacterial enzymes.[92] Monod was bold in his conjectures, some of which proved to be wrong or too simple, but in my view his efforts to find chemical explanations for complex biological phenomena represent a high point in the development of the modern biochemical sciences.

Among the less fruitful approaches to the problem of bacterial adaptation was that of the noted physical chemist Cyril Norman Hinshelwood, who, during the years after the Second World War, attempted to develop a kinetic theory of the "autosynthesis" of bacterial enzymes. In 1952 he stated:

> In the reproduction of a cell there is indeed copying, but degradation of energy also occurs, and highly complex series of reactions combine, as it were symphonically, to give among the total products certain substances of very low entropy. It is the interplay of all these processes which must make autosynthesis possible, not the replication of individual genes as such.[93]

Hinshelwood claimed that alterations in the enzymatic capacity of a bacterial culture are a consequence of a gradual transformation of most of the cells in response to changes in the nutrient medium. This interpretation was challenged by several investigators, notably Francis Ryan, who showed that such adaptation is a consequence of the spontaneous mutation of only a few cells which multiplied rapidly in the new medium.[94] Ryan's approach reflected the influence of the earlier invention, by Salvador Luria, of methods for determining the spontaneous mutation rate in a bacterial population. This "fluctuation test," for which Max Delbrück provided a mathematical formulation based on the so-called Poisson distribution (or Jacob Bernouilli's "law of large numbers"), represented a major step in the acceptance of the view that bacterial populations, like those of other

92. See Lwoff and Ullmann (1979).

93. Hinshelwood (1953), p. 1948. For a full account of Hinshelwood's microbiological research, see Dean and Hinshelwood (1966); see also Peacocke (1983), pp. 114–124.

94. Ryan (1952).

biological species, could be studied as genetic systems.[95] The demise of Hinshelwood's approach, no matter how soundly that approach may have been based on physical-chemical theory, offers, in my opinion, further evidence for the view that, without a knowledge of the organic-chemical constitution and biological properties of the organisms under study, the application of the most advanced theoretical knowledge of the time to the explanation of biological processes may not lead to further fruitful experimental inquiry.

In parallel with the studies, by non-medical microbiologists, of the adaptation of bacteria to changes in their chemical environment, some bacteriologists sought to explore the chemical composition of pathogenic microorganisms in the hope of discerning the cause of their infectivity. In these efforts, the studies of Oswald Theodore Avery and his associates on the pneumococcus occupy a unique place in the history of the biochemical sciences.[96] During the 1920s and 1930s, Avery's associates Michael Heidelberger and Walther Goebel had shown that the serological specificity of various known virulent strains of this organism lies in the chemical constitution of their capsular polysaccharides. Also, the English bacteriologist Fred Griffith had found in 1928 that the avirulent (non-encapsulated) strain could be "transformed" to the virulent strain by means of a heat-stable factor derived from dead virulent organisms. The subsequent sustained study in Avery's laboratory by James Alloway, Colin MacLeod, and Maclyn McCarty led, in 1944, to the publication of a paper which began as follows:

> Biologists have long attempted by chemical means to induce in higher organisms predictable and specific changes which thereafter could be transmitted in series as hereditary characters. Among microorganisms the most striking example of inheritable and specific alterations in cell structure and function that can be experimentally induced and are reproducible under well

95. Luria and Delbrück (1943). See also Luria (1984), pp. 73–79; in connection with this account, it is of interest that the Poisson distribution had been applied by Luria (1939) in a study of the effect of radiation on cultures of *Escherichia coli*. For Poisson, see Costabel (1978).

96. For Avery, see Dubos (1976) and McCarty (1985).

defined and adquately controlled conditions is the transformation of specific types of Pneumococcus.[97]

There followed the description, in meticulous detail, of the isolation of the active transforming principle and its chemical characterization as a "highly polymerized and viscous form of sodium desoxyribonucleate."[98] Although Avery did not claim in this paper the identity of the transforming factor with the genetic material of the pneumococcus, it later became evident from his private correspondence that he saw clearly the implication of the discovery for the problem he raised in that opening paragraph. Indeed, soon after the appearance of the 1944 paper, the noted geneticist Sewall Wright wrote: "The great possible significance of this observation in the interpretation of the role of nucleic acids of chromosomes and of other self-duplicating entities is obvious. . . . The results suggest the chemical isolation and transfer of a gene rather than induction of mutation."[99]

The significance of the identification of the pneumococcal transforming factor with DNA did not escape the attention of the biochemist Erwin Chargaff, who embarked on a program of research designed to determine quantitatively the chemical composition of DNA samples from a variety of biological organisms. By the use of the technique of partition chromatography, just introduced into protein chemistry by Martin and Synge, Chargaff's research group soon showed that the two purines (adenine and guanine) and the two pyrimidines (cytosine and thymine) in hydrolysates of such DNA samples are not present in equimolar proportions and arranged in the form of repeating tetranucleotide units, as had been suggested previously on the basis of the results obtained with less accurate analytical procedures.[100] There can be little doubt that the interpretation of the earlier data reflected the same kind of search for simplicity and regularity that afflicted some protein chemists before 1940. One of the reports from Chargaff's laboratory contained the following statement:

97. Avery et al. (1944), p. 137.
98. Ibid., p. 152. See Dunn (1969), pp. 61–64, for the relevant text of a letter (17 May 1943) from Avery to his brother, Roy Crowdy Avery.
99. Wright (1945), pp. 79–83.
100. Chargaff (1950, 1963).

> The results serve to disprove the tetranucleotide hypothesis. It is, however, noteworthy—whether this is more than accidental, cannot yet be said—that in all the deoxypentose nucleic acids examined thus far the molar ratios of total purines to total pyrimidines, and also of adenine to thymine and of guanine to cytosine were not far from 1.[101]

I will comment in the next chapter on some historical aspects of the reception accorded to the reports from Avery's and Chargaff's laboratories in relation to recent writings on the "origins of molecular biology."

Apart from the influence of the notable achievements sketched above, the 1940s also marked the demonstration, by Joshua Lederberg and Edward Tatum, of gene recombination in a strain of *Escherichia coli,* thus showing that a Mendelian theory of heredity is as applicable to bacteria (prokaryotes) as to higher plants and animals.[102] Also in the 1940s there began in the United States an intensive study of the bacterial viruses ("bacteriophages") by a tightly knit group of independent investigators influenced by Max Delbrück, who had been introduced to this field by Emory Ellis.[103] The "phages," as they soon came to be called, had been discovered in 1915–1917 by Frederick William Twort and Felix d'Hérelle, and for a time were considered to be potential therapeutic agents in the treatment of infectious diseases. Soon after their discovery, the geneticist Hermann Muller suggested that

> if these d'Hérelle bodies were really genes, fundamentally like our chromosome genes, they would give us an utterly new angle from which to attack the gene problem. They are filterable, to some extent isolable, can be handled in test tubes, and their properties, as shown by their effects on the bacteria, can then be studied after treatment. It would be very rash to call these bodies genes, and yet at present we must confess that there is no distinction known between the genes and them.[104]

101. Chargaff (1950), p. 206.
102. Lederberg and Tatum (1946); Lederberg (1987).
103. Ellis and Delbrück (1939).
104. Muller (1922), p. 48.

The prescience of this speculation may, in hindsight, be admirable, but at the time it was offered the phages were thought to be related to the filterable infectious agents found before in plants (for example, tobacco mosaic virus) and animals (for example, fowl pox virus). During the 1920s, however, the discovery of strains of *Escherichia coli* which, in the absence of free phage, could undergo lysis with the generation of phage particles indicated that there are significant differences among the so-called filterable viruses. One of the first investigators to seek an explanation of this phenomenon ("lysogeny") was Eugène Wollman of the Pasteur Institute. After his death during the Second World War at the hands of the Nazis, the problem was undertaken by André Lwoff, whose achievements confirmed and vastly extended Wollman's view that the phages are bearers of hereditary characters. I shall not attempt to summarize the subsequent outstanding discoveries made at the Pasteur Institute and various American laboratories except to note that, in its development after 1950, the study of phages became one of the principal arenas of the fruitful interplay of genetics and biochemistry, and was markedly advanced by the use of radioactive isotopes, enzyme purification, electron microscopy, and X-ray diffraction methods.[105]

Mention must be made, however, of the controversy that preceded and followed the report in 1935 by the organic chemist Wendell Stanley that he had isolated the tobacco mosaic virus (TMV) in the form of a crystalline globulin.[106] Stanley's analytical data for his preparation of TMV required correction by the plant pathologist Frederick Charles Bawden and the biochemist Norman Wingate Pirie, who showed TMV to contain measurable amounts of sulfur and phosphorus, and that in addition to protein it also contains a ribonucleic acid (RNA). Bawden and Pirie also gave some of their preparation to the crystallographers John Desmond Bernal and Isidor Fankuchen, who found that TMV possesses a liquid-crystalline structure.[107] Despite the shortcomings of Stanley's chemical characterization of the material he had isolated, his contribution was widely hailed, once again reflecting the appeal of "crystallinity" as a criterion of purity. Indeed, in 1946 he was cited by the Nobel Prize

105. Waterson and Wilkinson (1978).
106. Stanley (1935).
107. Bawden et al. (1936).

Committee for Chemistry for "his preparation of virus proteins in pure form." Ten years later, Gerhard Schramm and Heinz Fraenkel-Conrat showed independently that the infectivity of TMV resides in its RNA. I cannot forbear to note that the public acclaim accorded to Stanley was withheld from Avery before his death in 1955.

Apart from the problem of the chemical nature of the material isolated by Stanley, his report reactivated not only the debate as to whether the viruses were to be considered "living" or "non-living," but also the century-old argument about the unique properties which distinguish biological organisms from inanimate chemical objects.[108] During the course of the debate, viruses were said to be "living molecules" or "autocatalytic enzymes" and likened to genes or mitochondria. That argument subsided after massive evidence was presented for the view that the nucleic acid portion of a virus enters the genetic apparatus of the host cell. Moreover, the growing knowledge of the enzymatic complexity of that apparatus has rendered obsolete most of the attractive generalizations offered during the 1950s about the chemical mechanisms in the replication of genetic factors, in the biosynthesis and modification of proteins, in the regulation of intracellular processes, and many other aspects of the physiology of particular biological systems. New discoveries, whether they be the demonstration of the capacity of some RNA molecules to catalyze the cleavage of internucleotide bonds, or the finding of DNA-containing plant viruses and protein-free RNA viruses (viriods), have obliged biologists and chemists to revise long-held assumptions, and only strengthen the expectation that future discovery will require further change in biochemical thought and experimentation.

In the preceding pages, I have attempted to summarize briefly the remarkable growth of biological knowledge that occurred up to about 1950 as a result of the interplay of chemistry with the investigation of the morphology, physiology, and reproduction of living things. In dwelling on the controversies which attended that interplay, I sought to emphasize the limits placed at any time upon biological speculation by the available knowledge of the chemical constitution and interactions of the substances of which biological organisms were then known to be composed. In the remainder of

108. Pirie (1937, 1948); Wilkinson (1976).

this chapter I will comment on some features of the recent flurry of philosophical discourse about the interplay of biology and chemistry, against the background of the development and interactions of these two disciplines since about 1950. The prominent words or phrases in that discourse no longer include "vitalism" (as opposed to "mechanism"), both of which appear to have been disavowed by most of the participants; they seem to have been replaced by "organicism" (or "holism"), "anti-reductionism" (as opposed to "reductionism"), and "autonomy" (as opposed to "provincialism"), and some biologists prefer to speak of "teleonomy" rather than "teleology" (or "purpose"). Moreover, the ancient dispute about the difference beteen living and inanimate objects appears to have focused on two present-day bodies of biological thought—Darwinian natural selection as modified by Mendelian genetics and the study of the interaction of biological populations with their environment ("the evolutionary synthesis"), as opposed to a "molecular biology" based on the central role of nucleic acids in the transmission of hereditary characters and in the "expression" of phenotypic traits in the synthesis of proteins (especially enzymes) and the regulation of processes in which such proteins are involved. Much has been written about the difference in biological outlook and methodology between those scientists who seek to define, through the study of morphological and physiological change during the course of biological evolution, such concepts as "fitness" or "adaptation" in the interaction of "species" with one another and their natural environment, and those scientists who seek to explain biological phenomena in terms of such concepts as the "translation of a genetic code" or "complementarity of chemical interaction" or the role of "allostery" in the regulation of some physiological processes through changes in the "conformation" of protein molecules. Less has been said about the problem of embryonic development, and there has been a tendency to take the growth of biochemical knowledge for granted or, what has been more common, to exhibit ignorance of (or willful indifference to) the historical role of chemical thought and experimentation in the search for explanations of biological phenomena. The principal participants in the recent debate have been members of a group of specialists in the philosophy of biology, as well as several prominent evolutionary biologists, molecular biologists, and physicists.[109]

109. The number of books published since about 1965 by professional philoso-

A "Sack Full of Enzymes"?

One attitude about the place of biochemistry among the biological sciences was expressed in 1949 by Max Delbrück, who had turned from theoretical physics to phage genetics:

> Biology is a very interesting field to enter for anyone, by the vastness of its structure and the extraordinary variety of strange facts it has collected, but to the physicist it is also a depressing subject, because, insofar as physical explanations of seemingly physical phenomena go, like excitation, or chromosome movements, or replication, the analysis seems to have stalled around in a semidescriptive manner without noticeably progressing towards a radical physical explanation. He may be told that the only access of atomic physics to biology is through biochemistry. Listening to the story of modern biochemistry he might become persuaded that the cell is a sack full of enzymes acting on substrates converting them through various intermediate stages either into cell substance or into waste products. The enzymes must be situated in their proper strategic positions to perform their duties in a well regulated fashion. They in turn must be synthesized and must be brought into position by manoeuvres which are not yet understood, but which, at first sight at least, do not necessarily seem to differ in nature with the rest of biochemistry. Indeed, the vista of the biochemist is one with an infinite horizon. And yet, this program of explaining the simple through the complex smacks suspiciously of the program of explaining atoms in terms of complex mechanical models. It looks sane until paradoxes crop up and come into sharper focus. In biology we are not yet at the point where we are presented with clear paradoxes and this will not happen until the analysis of the behaviour of living cells has been carried into far greater detail. This analysis should be done on the living cell's own terms

---

phers of biology is now considerable; I have consulted those by Simon (1971), Ruse (1973), Hull (1974), Grene (1974), and Rosenberg (1985). Books containing essays by these and other philosophers include those edited by Grene and Mendelsohn (1976) and by Ruse (1989). In recent years, philosophical problems have been of special interest to evolutionary biologists, notably Mayr (1982, 1988), and several molecular biologists, among them Crick (1966), Monod (1970), Jacob (1970, 1982), and Delbrück (1986).

> and the theories should be formulated without fear of contradicting molecular physics. I believe that it is in this direction physicists will show the greatest zeal and will create a new intellectual approach to biology which would lend meaning to the ill-used term biophysics.[110]

This provocative statement by a scientist who has been described by his acolytes as a principal figure in the "origins of molecular biology"[111] may be taken as a starting point for the consideration of some aspects of the recent philosophical discourse about the relation of biology to chemistry.

First, as regards the rhetorical comment that, to a physicist, the living cell might have appeared to be, in 1949, a "sack full of enzymes," I believe it is correct to say that no nineteenth-century or twentieth-century biochemist who worked on enzymes ever described a living cell in this way. Indeed, men such as Franz Hofmeister, Frederick Gowland Hopkins, and (later) Hans Krebs repeatedly emphasized the importance of the problem of the intracellular organization of the spatial arrangement and translocation of the chemical constituents of a cell, and the regulation of the catalytic activity of its enzymes.[112] Each of these leading biochemists, representing three successive generations, saw this problem in the light of the cytological and chemical knowledge of their day, but did not advocate, as did other biologists, encouraged by the "organismic" philosophy of Alfred North Whitehead, a "neo-vitalistic" attitude in biology. Between 1900 and 1940, methods were developed for the isolation and study of some enzymes from disrupted cells, but reliable knowledge about the localization of such enzymes within the formed elements of cells came only after the availabilty of instruments (for example, the preparative ultracentrifuge) for the physical separation of entities such as mitochondria. These organelles were then shown by cytologists (largely by means of improved methods of electron microscopy) and biochemists to be complex organized structures in which are located the large set of previously identified enzymes involved in the oxidation of metabolites by molecular oxygen, with the utilization of the free energy derived from such processes for the synthe-

110. Delbrück (1949), p. 191.
111. Cairns et al. (1966); Stent (1968).
112. Hofmeister (1901); Peters (1930); Hopkins (1933); Krebs (1971).

sis of adenosine triphosphate (ATP) from adenosine diphosphate (ADP) and inorganic phosphate. Also, the study of the cytoplasmic structures (ribosomes) involved in the synthesis of proteins has shown them to be complex organized assemblies of units which include key enzymes in the formation of peptide bonds. Such complexity of intracellular organization, which has been made evident by the continued interplay of chemical and cytological investigation and whose details are still under active study, may not be intellectually satisfying to a physicist or a philosopher who wishes to seek elegantly simple non-chemical "laws" which distinguish living things from inanimate ones. Thus far, however, words of "learned depth and thundering sound" have been largely irrelevant to the day-to-day efforts of chemists and biologists of various kinds to describe, as accurately and as fully as the available methods will permit, the organization of biological systems, and to devise or use any new methods which may help to enlarge understanding of the relationship between biological "wholes" and their biological and chemical "parts."

The "bag of enzymes" caricature of biochemical research was echoed in 1953 by Boris Ephrussi: "Trained in the tradition of the theory of solutions, many a biochemist tends, even today, to regard the cell as a 'bag of enzymes.' However, everyone realizes now that the biochemical processes studied *in vitro* may have only a remote resemblance to the events actually occurring in the living cell."[113] As an advocate of the importance of cytoplasmic inheritance, Ephrussi also noted the limited value of genetics in the study of the "processes of cell integration."

Certainly, long before 1953, biochemists had stressed the fact that many intracellular enzymes, especially those concerned with oxidative and synthetic processes, appeared to have been inactivated upon disruption of cell structure, and called attention to the need to consider the "organization," "integration," and "regulation" of enzyme systems.[114] These views were expressed in various ways, with a preponderant emphasis on the colloid chemistry of protoplasm. Moreover, there had been considerable discussion of the concept of "integrative levels" in biology. For example, Novikoff wrote:

113. Ephrussi (1953), p. 108.

114. Some of these views have been summarized by Peters (1930), Needham (1936), Korr (1939), and especially by Teich (1973).

> Since higher level phenomena always include phenomena at lower levels, one can not fully understand the higher levels without an understanding of the lower levels as well. But a knowledge of the lower levels does not enable us to predict, *a priori*, what will occur at a higher level . . . The concept of integrative levels indicates to research biologists the crucial aspects of their problems, the solution of which puts the known facts into proper perspective by revealing the decisive element, the element imparting the uniqueness to the phenomena under study. It emphasizes the importance of studying the "mesoforms," matter at the point of transition from one level of organization to the next, so as to deepen our understanding of the unique qualities of the higher level. For example, it would indicate that an intensive study of the transition region between the chemical and biological levels, between protein and protoplasm, will help reveal the organizing relations unique to living matter and fundamental to vital activities.[115]

This quotation may be compared with one taken from a book by François Jacob, published twenty-five years later, after the emergence of "molecular biology" from the fruitful interplay of chemistry and genetics during that period:

> Organisms build themselves by means of a series of integrations. Similar elements assemble themselves into an intermediate ensemble. Several of these ensembles associate with each other to constitute an ensemble at a higher level and so on. It is therefore by the combination of more and more intricate elements, by an articulation of structures which are subordinate to one another, that the complexity of living systems is born. And if at each generation these systems can reproduce themselves from their elements, it is because at each level the intermediate structure is thermodynamically stable. Living things thus construct themselves by means of a series of packagings. They are arranged according to a hierarchy of discontinuous ensembles. At each level, the units of relatively well defined size and nearly identical structure unite to form a unit at a higher level.

115. Novikoff (1945), pp. 214–215. For a biography of Novikoff, see D. R. Holmes (1989).

> Each of these units formed by the integration of the subunits may be denoted by the general term *integron*. An integron forms itself by the assembly of integrons of a lower level; it participates in the construction of an integron of a higher level.[116]

Although Jacob's statement is based on vastly more chemical and biological knowledge than was available to Novikoff, it may be questioned whether the introduction of the word *integron* markedly alters the philosophical status of the experimental problems which have challenged embryologists and cytologists, as well as those investigators who have sought, in part successfully, to define some of the links between the chemical structure and interactions of individual molecular species and their roles in the organized, integrated, and regulated processes associated with the phenomena exhibited by living things.

An aspect of "organicism" which, in my opinion, merits consideration is implicit in the statement by the noted evolutionary biologist Ernst Mayr that "even at the molecular level, the macromolecules that characterize living beings do not differ in principle from the lower-molecular-weight molecules that are the regular constituents of inanimate nature, but they are much larger and more complex. This complexity endows them with extraordinary properties not found in inert matter."[117] I am certain that in these sentences Mayr did not intend to revive the nineteenth-century "energy-rich" proteins of Ludimar Hermann, Eduard Pflüger, or Max Verworn, but I found an echo of Whitehead's statement that

> the relation of part to whole has the special reciprocity associated with the notion of organism, in which the part is for the whole; but this relation reigns throughout nature and does not start with the special case of the higher organisms. Further, viewing the question as a matter of chemistry, there is no need to construe the actions of each molecule in a living body by its exclusive particular reference to the pattern of the complete living organism. It is true that each molecule is affected by the aspect of this pattern as mirrored in it, so as to be otherwise than what it would have been if placed elsewhere. . . . The mode of

116. Jacob (1970), p. 323. This translation differs from that in Jacob (1973), p. 302.
117. Mayr (1988), p. 14.

approach to the problem, so far as science is concerned, is merely to ask if molecules exhibit in living bodies properties which are not to be observed amid inorganic surroundings.[118]

In relation to Mayr's statement, a skeptical biochemist must note that a macromolecule such as hemoglobin, whether in a crystalline state or dissolved in a suitable artificial medium, is just as "inert" as any other lower-molecular-weight constituent of biological systems, for example insulin, epinephrine, or thiamine. Whitehead's statement raises the question whether, as a component of an "organized" mammalian erythrocyte, hemoglobin exhibits any "extraordinary" properties not observable in experiments with the isolated protein.

Among the physiological processes in higher animals, breathing and the role of hemoglobin in this process have, for at least a century, presented some of the most challenging problems in the interplay of biology and chemistry. The story cannot be adequately summarized here, except to note that the long list of distinguished participants in the experimental study of these problems before 1940 includes the names of such physiologists as Carl Ludwig, Christian Bohr, John Scott Haldane, and Joseph Barcroft, as well as the chemists Felix Hoppe-Seyler, Hans Fischer, Lawrence Joseph Henderson, and Donald Dexter Van Slyke.[119] If, among the chemists, the leading place is now assigned—and I believe rightly—to Max Perutz, it is because through his sustained X-ray studies on the detailed three-dimensional structure of hemoglobin crystals he was able to explain, in chemical terms, phenomena which had been previously described by physiologists and biochemists.[120] As far as hemoglobin is concerned, therefore, when it is present in the red cells of the mammalian blood circulation, its *chemical* properties appear to be no different from those observed for the isolated protein. Thus, the physiological function of hemoglobin in breathing has, at least in great part, been "explained" in terms of its detailed chemical structure and the changes in that structure when hemoglobin interacts with the other chemical substances it encounters in the blood. By providing such an explanation, Perutz has enlarged the scope of experimental inquiry into the relation between the chemical struc-

118. Whitehead (1925), pp. 214–215.
119. Edsall (1972).
120. Perutz (1970).

ture of proteins and their functional roles as parts of living organisms.

On Biomolecular Structure

At this point I find it necessary to insert a digression regarding the use of the word "structure" in discourse about chemical molecules and biological organisms. In organic chemistry, the nineteenth-century problem of the organization of chemical atoms in chemical molecules was partially solved, after much controversy, by the adoption during the 1860s of the idea that the atoms are held together by "valence bonds" to form three-dimensional assemblies which were considered to represent the "structure" of a particular molecular species. In the background of the controversy was a large body of empirical knowledge about the reactions of various carbon compounds, and of theories designed to unify and simplify the seemingly growing complexity of organic chemistry in terms of such concepts as "radicals" or "types." One of the objectives of these theoretical efforts was to develop a taxonomy for the classification of chemical substances.[121] As in biology since Aristotle, Linnaeus, and Cuvier, and in chemistry since Lavoisier, during the first part of the nineteenth century taxonomy was considered to be an important part of organic-chemical theory. It still remains an essential component of biological theory and practice,[122] but after the adoption of Kekulé's system of classification, based on the structure theory, that system ceased to be a taxonomic tool, in large part because its use became the basis for the successful synthesis of new organic compounds.[123] This success, with great impact on the development of a large-scale chemical industry and the creation of new drugs for use in medical practice, gave support for the practical utility of the "metaphysical" concepts of atom, molecule, and valence employed in the elaboration of new methods in organic chemistry. For the pur-

121. See Freund (1904); Russell (1971); Fisher (1973b, 1974); Rocke (1984).

122. Mayr (1982), pp. 140–297. It should be recalled that before about 1800, much of what we now call "chemistry" was considered by many scholars to be the "inorganic" and "organic" parts of "natural history" and distinct from "natural philosophy."

123. The first edition of Friedrich Beilstein's *Handbuch der organischen Chemie*, based on Kekulé's classification and published during 1880–1883, listed about 15,000 compounds. By the 1980s this number had risen to more than three million known organic compounds.

poses of this chapter, it is noteworthy that conjectures about the structures of natural organic compounds were either proved or disproved by the rational synthesis of such "wholes" from reactive derivatives of their "parts." For example, in 1875 Ludwig Medicus proposed a Kekulé-type structure for uric acid, which appeared to be related to xanthine, caffeine, and theobromine, but the status of his claim was uncertain because some of the experimental findings upon which it was based could not be confirmed. In 1882 Emil Fischer, who initially questioned the validity of Medicus's proposal, began a systematic synthetic investigation of the structure of these compounds and not only showed that Medicus had been right, but also demonstrated that all these compounds (as well as adenine and guanine) are derivatives of a bicyclic system which Fischer named "purine."[124] Fischer entered the field of protein chemistry in 1899 with the objective of synthesizing a protein, and during the succeeding decade he developed a method for the synthesis of polypeptides.[125] Although the massive effort of his research group laid the foundations of peptide chemistry, Fischer failed to repeat his earlier notable successes in the fields of purine and carbohydrate chemistry. In claiming that some of the polypeptides synthesized in his laboratory had the properties of the ill-defined "albumoses" and "peptones" produced upon the cleavage of proteins by enzymes such as pepsin, Fischer chose to dismiss the evidence offered by biochemists and physical chemists for the relatively large molecular size of the proteins.

Thus, from the point of view of a modern organic chemist, the validity of a hypothesis about the structural arrangement of the atoms in a chemical compound derived from a biological source rests upon the synthesis of that compound from smaller molecules of known structure by means of a procedure which is consistent with the conceptual framework of organic chemistry at that time. If the properties of the synthetic product are the same as those of the biochemical substance in all measurable respects (within the precision of the methods and instruments which are used), the chemical structure of the natural compound is considered to have been established. In principle, if not in practice, it does not matter whether the hypothesis was derived from the use of the relatively limited analytical

124. Fischer (1907).
125. Fischer (1906, 1923); see Fruton (1990), pp. 185–196.

methods available around 1900 or from the present-day reliance on such sophisticated physical methods as two-dimensional nuclear magnetic resonance (NMR) spectroscopy, mass spectrometry, X-ray diffraction analysis, Fourier-transform infrared spectroscopy, or computerized versions of older instruments. Because analytical practice today is based on a vastly greater body of theoretical and experimental knowledge in both physics and chemistry, the formulation of a proposed structure is now less challenging than it was during the first decades of this century. As discussed in the preceding chapter, in the field of protein chemistry Sanger's determination of the amino acid sequence in insulin involved the use of new chemical reagents and of enzymes of known specificity, as well as the availability of new methods for the separation of peptides.[126] In the determination of the nucleotide sequence in preparations of DNA, the two ingenious chemical methods introduced in 1977 depend on the use of radioactive phosphorus (^{32}P) as a label and of newly discovered enzymes which act on DNA.[127] It should be added that, at present, the detailed structural interpretation of X-ray diffraction data for protein crystals still depends on prior reliable knowledge of the amino acid sequence of the peptide chain (or chains) in the protein under study.[128]

During the past four decades there have also been remarkable advances in the art of organic-chemical synthesis of complex compounds of biological origin. Few areas of chemical investigation require as much imaginative talent, as well as the collective efforts of large, closely directed research groups. In meeting the challenges presented by the synthesis of such molecules as the steroid hormones or chlorophyll, organic chemists developed chemical theory through refinement of older concepts, such as those of chirality and

126. In particular, the chemical degradation method devised by Pehr Edman during the 1950s has provided the basis for an automatic apparatus ("sequenator") for the determination of the amino acid sequence of long-chain polypeptides.

127. Maxam and Gilbert (1977); Sanger et al. (1977).

128. Abir-Am (1982a, p. 305), a prolific writer about biochemistry, chose to limit the use of the word "structure," as applied to proteins, to "three-dimensional structure." In this eccentricity she disregarded the convention according to which the linear sequence of amino acid units in a peptide chain is termed "primary structure," the stabilization of such a chain through hydrogen-bonding is termed "secondary structure," the folding of such stabilized chains is known as "tertiary structure," and the association of separate folded chains in oligomeric proteins as "quaternary structure."

conformational analysis, well as the quantum-mechanical treatment of valency. They also invented new chemical methods for the control of the stereospecificity of synthetic reactions, and made ingenious use of metal ions and enzymes as catalysts. Among the many achievements of the recent past, that of Robert Woodward and Albert Eschenmoser, and their respective research groups, in the total synthesis of vitamin B_{12} (cyanocobalamin) is especially memorable.[129]

Another facet of the recent development of organic-chemical synthesis came from the recognition that many hormones, like insulin, are peptides, thus providing a powerful stimulus to the invention of new methods in the field opened by Emil Fischer. Not only were the structures of many peptide hormones established by stepwise synthesis from reactive amino acid derivatives, with isolation and identification of the intermediate peptides, but systematic studies were also conducted on the effect of changes in the structure of such hormones on their physiological activity. Such studies, in turn, provided a chemical basis for efforts to identify the nature of the interactions of the hormones with their cellular "receptors." Peptides of moderate length (15–20 amino acid units) can now be made reliably in a programmed automatic "synthesizer" by modifications of a "solid-phase" method invented by Bruce Merrifield. An analogous instrument has been developed for the synthesis of oligonucleotides. It should be noted, however, that since each step in such an automated procedure cannot be expected to be 100 percent effective, the final product is likely to contain compounds other than the one desired, and extensive purification by chromatographic methods may be necessary. Consequently, if the final yield of a compound having the postulated structure is relatively low, questions may be raised about the validity of the claim that the structure has been confirmed by synthesis.

It is my impression that many biochemists, and perhaps some chemists as well, consider the present-day chemical and physical methods of analysis, together with X-ray crystallography, to be sufficiently precise to establish unequivocally the three-dimensional structure of such complex substances as proteins and nucleic acids,

129. Eschenmoser and Wintner (1977). Another notable achievement of the recent past is that of Gobind Khorana's group in the synthesis of oligodeoxyribonucleotides (Caruthers, 1985).

and that the elegant but laborious efforts of the research groups associated with chemists such as Woodward represented a form of artistic expression on the part of the group leader, rather than a necessary part of the establishment of the validity of a proposed molecular structure. There is no doubt that the chemical structure of proteins (or nucleic acids) having a molecular weight greater than about 6,000 is more complex than that of such substances as chlorophyll or cyanocobalamin, and that a description of their structure must include features which are less prominent in smaller complex molecules. Thus, in addition to the concepts of the covalent bond (the sharing of the electronic orbitals of atoms) and bond direction, distances, and angles, one must consider such non-covalent weaker interactions as the formation of "hydrogen bonds" (as between the oxygen of a CO group of an amino acid unit in one part of a peptide chain and an NH group of another amino acid unit in another part of the chain). Moreover, in some regions in the interior of a protein molecule, "ionic bonds" (formed by the attraction of carboxylate and immonium ions) and the "hydrophobic attraction" of water-insoluble parts of amino acid units also have been shown to play significant roles in the manner in which a polypeptide chain assumes the form characteristic of "globular" proteins. If to these features of protein structure one adds the so-called van der Waals nonspecific interactions, it is clear that the questions presented by the three-dimensional chemical structure of proteins and nucleic acids still represent important areas of inquiry, and that the more exact definition of such non-covalent interactions presents a challenge to chemists interested in biological problems.

During the 1950s, Christian Anfinsen and his associates demonstrated that after suitable inactivation of the enzyme ribonuclease A, this protein (a single polypeptide chain of 124 amino acid units with four disulfide bridges) could be made to regain almost all of its original catalytic activity. Extensive evidence was provided for the conclusion that the inactivation procedure (which involved reduction of the disulfide groups) led to the disorganization of the specific three-dimensional structure of the protein, and that the catalytic activity and crystal structure were almost fully regained by controlled oxidation to regenerate the disulfide groups.[130] These important findings supported the view that the organization of the structure of

130. See Anfinsen and Haber (1961) and references therein to previous work.

this enzyme is determined solely by the sequence of the amino acid units in its polypeptide chain. It is now widely accepted that the specific "folding" of a polypeptide chain made in a biological system is entirely a chemical consequence of its amino acid sequence, as determined by the sequence of nucleotides in the DNA of that biological organism. However, the fact that many protein chemists are still engaged today in the study of the "folding problem" suggests that more can be learned about the factors which determine the three-dimensional structure of proteins.[131]

A significant outcome of the biochemical knowledge gained recently about proteins is that many of these macromolecules contain a relatively small assembly of amino acid units associated with the biological functions of a particular protein. Thus, we now speak of "active sites" of enzymes or "receptor sites" of regulatory proteins, and the view has emerged that the large size of a protein confers stability to a structure in which there are flexible regions which interact specifically with smaller molecules or with parts of other macromolecules, or which can undergo structural change in response to such interactions at relatively distant "binding sites" in a protein.

These conclusions have not gone unnoticed by philosophers of biology. For example, after an excellent description of Perutz's findings, Alexander Rosenberg wrote:

> The two key theoretical achievements were Linus Pauling's theory of the chemical bond, which enables biochemists to specify a three-dimensional secondary structure given a one-dimensional primary structure; and Monod's appreciation that biological macromolecules, especially enzymes, and proteins engaged in transport, can and do reversibly change their molecular shape as a consequence of interactions with other molecules, and that these changes in shape determine their effects—regulate their functions. The importance of this allosterism cannot be overemphasized: The existence of this phenomenon provides molecular biologists with a tool of the greatest power in their attempts to explain the biological in terms of the chemical.[132]

131. See, for example, Fasman (1989).
132. Rosenberg (1985), p. 81.

The influence of the theoretical contributions of Pauling and of Monod cannot be questioned, but some reservations must be noted about this sweeping statement. As regards Pauling, in 1951 he published a set of eight key papers on the structure of proteins. The first of these papers began as follows:

> During the past fifteen years we have been attacking the problem of the structure of proteins in several ways. One of these ways is the complete and accurate determination of the crystal structure of amino acids, peptides, and other simple substances related to proteins, in order that information about interatomic distances, bond angles, and other configurational parameters might be obtained that would permit the reliable prediction of reasonable configurations for the polypeptide chain. We have now used this information to construct two reasonable hydrogen-bonded helical configurations for the polypeptide chain; we think it is likely that these configurations constitute an important part of the structure of both fibrous and globular proteins, as well as of synthetic polypeptides.[133]

In that paper, the authors described two structures which they termed the "3.7-residue helix" (later the alpha-helix) and the "5.1-residue helix," and they also criticized earlier hydrogen-bonded helical models proposed by Maurice Huggins in 1943 and by Bragg, Perutz, and Kendrew in 1950. In another paper in the series, Pauling and Corey described a "pleated-sheet configuration," with hydrogen bonding between adjacent polypeptide chains. The formulation of these models involved the assumption, based on the theory of resonance and X-ray diffraction data for a few amides, that all the atoms of a -CO-NH- group lie in a single plane. Some years later Pauling described how he came to discover the alpha-helix in 1948, while recuperating from a cold in Oxford:

> It occurred to me to make a search for the simplest configurations—those in which all of the amino-acid residues are structurally equivalent. The most general operation that con-

133. Pauling et al. (1951), p. 205. See Olby (1974), pp. 267–295. Helical structures had been proposed by Charles Hanes (1937) and by Robert Eugene Rundle and Dexter French (1943) for the long polyglucosyl chains of the amylose fraction of starch.

> verts an asymmetric element into an identically equivalent element (not its mirror image) is a rotation about an axis combined with a translation along the axis. The repetition of this general operation automatically leads to a helix. I attempted accordingly to find helical configurations of polypeptide chains involving planar amide groups with known dimensions, such that suitable hydrogen bonds were formed. Within an hour, with the aid of a pencil and a piece of paper, I had discovered a satisfactory helical structure. It did not, however, explain the details of the x-ray diagram of hair and other alpha-keratin proteins, and nothing more was done along these lines for some months. After my return to Pasadena, Professor Corey and I suggested to Dr. H. R. Branson, who was interested in the application of mathematics to chemical problems, that he make a search for other satisfactory helical configurations. He found only one, and in 1951 a description of the two helices . . . was published.[134]

Pauling gave this account in a lecture on the application in chemistry of the "stochastic" method (from the Greek *stokhastikos*, which Pauling defined as "apt to divine the truth by conjecture").

As it turned out, in 1959 John Kendrew showed that sperm whale myoglobin gave an X-ray diffraction pattern which indicated that about 75 percent of the single polypeptide chain of this protein is in the form of rod-like alpha-helices. As more crystalline proteins were examined in succeeding years, however, many of them showed much less helical content, and some, such as alpha-chymotrypsin, do not appear to contain any detectable alpha-helices. The so-called 5.1 residue helix, although thought by Pauling to be a "satisfactory" model, has been removed from consideration as an element of protein structure, and the X-ray data for keratin which appeared to support this hypothesis were reinterpreted as indicating coils of alpha-helices. The pleated-sheet structure has been identified in parts of many proteins, and is especially prominent in silk fibroin, but there are also stretches of the peptide chain in the form of irregular "turns." Features of protein structure not predicted by the "stochastic method," but found to be prominent in many proteins,

134. Pauling (1955), p. 294. A biography of Pauling was prepared by Serafini (1989); for critical reviews, see Roberts (1990) and Kay (1990).

include "hairpin bends" where a polypeptide chain makes a sharp turn by the formation of hydrogen bonds linking amide groups separated by three amino acid units. Also, the presence in the interior of some globular proteins of a high proportion of the amino acid units bearing water-insoluble side chains indicates that hydrophobic interactions play a significant role in the assembly of the three-dimensional structure of such proteins. Moreover, many single-chain proteins are composed of "domains" (assemblies of aggregated portions of the polypeptide chain) which are linked by short peptide segments, thus forming a "cleft." Such a cleft has been shown, in several cases, to be the location of the active site of an enzyme or of a receptor protein.[135]

This brief summary of the present state of the problem that Pauling attacked by the stochastic method may perhaps suffice to suggest that the fears of some biologically-minded philosophers or philosophically-minded biologists that chemistry may soon explain all of biology are somewhat premature. As in the study of the organization of biological systems, so in the study of protein structure seeming simplicity has given way to complexity in the face of the empirical knowledge that has been provided since about 1960 by X-ray crystallography aided by modern computer technology. Although this method has been indisputably the most powerful experimental approach to the determination of the intimate chemical structure of protein molecules, it should also be recalled that no method, however powerful, is free of limitations. A skeptical biochemist may perhaps still ask whether a protein molecule has the same conformation, especially of its active site units, in the crystalline state as it does in the dissolved state or as part of an intracellular apparatus. The kind of question he might raise is what changes, if any, in conformation arise from lattice forces within a protein crystal. He might also ask whether the intracellular organization of the three-dimensional structure of a globular protein is an automatic consequence of the amino acid sequence of its polypeptide chain (or chains), especially in view of the recent appearance on the biochemical scene of "chaperone" proteins which accompany newly created polypeptide chains and may play a role, not determined at

135. For an example of an enzyme see James and Sielecki (1983), and for one of a receptor protein see Quicho (1990).

the present writing, in folding such chains into conformations that might not have been predicted from the sequence of their amino acids.

The most important limitation of X-ray crystallography is that although the use of this method has provided evidence of changes in the conformation of parts of a protein when it binds a chemical compound, it cannot at present tell anything about the dynamics of such processes. It is now acknowledged that in order to perform their biological function, whatever that might be, some parts of a protein must be sufficiently flexible to perform that function and then to return to their previous conformational state. In this connection, it should be noted that recent advances in nuclear magnetic resonance (NMR) spectroscopy have made it possible to determine the three-dimensional structure of relatively small proteins (molecular weights up to about 12,000) in aqueous solution. The "disorder" at the molecular surface of the structures derived from such NMR studies appears to be more pronounced than that for the corresponding structures derived from X-ray studies. The study of such oscillatory dynamics is, in my opinion, one of the most challenging tasks in the continuation of the effort to link protein structure to biological function, and requires new methods (and improvement of older ones) which will permit the accurate measurement of fast reactions in the 10^{-12} to 10^{-15} second range.[136]

Jacques Monod and "Allostery"

Jacques Monod's approach to biological problems was very different from that of Pauling. Whereas Pauling sought to apply his physical-chemical theories of valence and structure to various problems suggested by biologists, Monod's theories came largely from his sustained personal research on the biological problem of bacterial growth.[137] In particular, he set himself the task of "evaluating the respective role of hereditary factors (i.e. genes or other self-duplicating units) and environmental factors (substrate) in the synthesis of

136. For a summary of recent NMR studies on protein structure, see Wüthrich (1990). See also Brandén and Jones (1990) and Janin (1990) for critical discussions of the precision of some of the protein three-dimensional structures derived from X-ray diffraction and NMR data. For reports on studies of fast reactions in biochemical systems, see Clementi et al. (1985) and Williams (1987).

137. See Lwoff (1977).

an enzyme."[138] From his studies on the utilization of various sugars in the "diauxic" growth of *Escherichia coli* cultures, Monod recognized the importance of the enzyme beta-galactosidase. By the use of mutant strains of *E. coli,* he and his associates demonstrated that what had been called "enzymatic adaptation" is a process controlled by a closely linked set of genes (an "operon"). This process includes not only the "induction" and "repression" of the activity of genes in the synthesis of beta-galactosidase, but also the participation of a transport system (which Monod named "permease") for the entry of sugars into the bacterial cell. Moreover, in order to explain the transmission of the genetic "message" to the cytoplasmic sites of protein synthesis, Monod and his associate Jacob postulated the existence of a short-lived "messenger RNA." These concepts, and the experimental data on which they were based, were summarized in a review article (1961) which, in my opinion, is one of the most significant (and most beautifully composed) contributions to modern biochemical thought.[139] By the time this article was written, it had been widely accepted that the word "gene" denotes a nucleotide sequence in DNA, and the model proposed for the lactose operon consisted of an arrangement of its nucleotides in the following order: regulator gene, promoter site, operator site, and "structural" genes for beta-galactosidase, an acetylase, and the permease. Some aspects of the theory, as presented in 1961, later required revision.

Monod's continued interest in the control and regulation of cellular metabolism found expression during the 1960s in a series of papers in which he and his colleagues developed a theoretical model designed to explain a particular property of some enzymes. It had been known for a long time that when, in an aqueous solution, an enzyme such as invertase catalyzes the hydrolysis of the glycosidic bond in a suitable substrate (whose total molar concentration [S] is much greater than that of the enzyme [E]), a series of replicate determinations of the initial velocity (v) of the reaction at increasing values of [S] provides data for a graphical plot of v versus [S] in the form of a rectangular hyperbola, or of $1/v$ versus $1/[S]$ in the form of a straight line. From such plots one may estimate, with fair precision, the magnitude of two parameters: the so-called Michaelis constant (K_M) and the maximal velocity ($V = k_{cat}[E]$). This kind of calcu-

138. Monod (1947), p. 261.
139. Jacob and Monod (1961). For "messenger RNA" see Brenner et al. (1961).

lation rests on at least two assumptions: (1) that the initial step in the catalysis is the "productive" reversible binding of the substrate by the enzyme, and (2) that the measured maximal velocity is that attained when the catalytic region of all the enzyme molecules is bound to substrate molecules. These assumptions imply that the explanation of the efficiency and specificity of enzymes such as invertase resides in the magnitude of the equilibrium constants for the "productive" binding of substrates at their catalytic regions. From the important studies of Myron Bender and other chemists, we now know that these assumptions must be tested for each enzyme under study, and may be invalid in the case of the action of the same enzyme on closely related compounds, such as the amide and ester substrates of the enzyme chymotrypsin.[140]

The particular property of some enzymes identified during the 1950s as participants in metabolic pathways, and which served as the stimulus to Monod's enunciation of his theory of "allostery," appears to have been the finding of his doctoral student Jean-Pierre Changeux that plots of initial velocity versus substrate concentration for the action of the threonine deaminase of *E. coli* did not give rectangular, but sigmoidal, curves. Such behavior had been noted earlier during the 1950s by Edwin Umbarger and by Richard Yates and Arthur Pardee for other isolated bacterial enzymes involved in some biosynthetic pathways, and had been considered to be important in their intracellular regulation through "negative feedback" inhibition of an early step in these pathways. A formal analogy to such sigmoidal responses was provided by the observation in 1904 by Christian Bohr and his associates that the reversible binding of oxygen by hemoglobin is markedly affected by the concentration of carbon dioxide in the blood, with an accentuation of the S-shaped plots of the extent of oxygenation (expressed as a percentage of the maximum) against the oxygen pressure. The importance of this phenomenon in the physiological process of breathing was soon recognized, especially through the studies of Joseph Barcroft and Lawrence Joseph Henderson.[141] Chemical explanations of the "Bohr effect" came later from the work of many investigators on the interaction of hemoglobin with hydrogen ions and carbon dioxide, and most recently

140. Bender and Kezdy (1965).

141. Barcroft (1928); Henderson (1928). See also Edsall (1972, 1986) and Dickerson and Geis (1983).

in the sustained X-ray crystallographic studies of Max Perutz. In hindsight, importance is now attached to the equation proposed in 1910 and 1913 by Archibald Vivian Hill, who explained the sigmoidal oxygenation curve by means of the assumption that the process involves the combination of *n* molecules of oxygen with "aggregates" of hemoglobin molecules. Thus, for human hemoglobin, *n* (the so-called Hill coefficient) was taken to be about 2.6. At that time, it had not yet been established that hemoglobin is an oligomeric protein, composed of four subunits each of which binds oxygen at its iron-porphyrin (heme), and even during the 1920s when this feature of the hemoglobin molecule was known, the Hill coefficient had no chemical meaning except to imply that the four hemes did not act independently of one another in the oxygenation reaction.[142] The recognition that the Bohr effect is related to the sequential and cooperative binding of the four oxygen molecules to hemoglobin, and that this phenomenon is associated with a change in the conformation of the protein, was expressed in 1951 by Jeffries Wyman, in a paper with David West Allen: "The reason why certain acid groups are affected by oxygenation is simply the alteration in their position and environment which results from the change of configuration of the hemoglobin molecule as a whole accompanying oxygenation."[143] In that paper, it was also suggested that "if we are prepared to accept hemoglobin as an enzyme, its behavior might give us a hint as to the kind of process to be looked for in enzymes generally."[144]

It should be noted that in the first extended statement of the theory of "allostery" by Monod, Changeux, and Jacob, the emphasis was primarily on its relevance to the regulation of enzyme action and that, although they mentioned the analogy to the oxygenation of hemoglobin, the discussion in that paper was based on a

> . . . general model schematizing the functional structures of controlling proteins. These proteins are assumed to have two, or at least two, stereospecifically different, non-overlapping receptor sites. One of these, the *active site*, binds the substrate and is responsible for the biological activity of the protein. The other, or

142. Wyman (1949), p. 101.
143. Wyman and Allen (1951).
144. Ibid., p. 515. This and the previous quotation are from Edsall (1985); see also Edsall (1990).

allosteric site, is complementary to the structure of another metabolite, the *allosteric effector*, which it binds specifically and reversibly. The formation of the enzyme-allosteric effector complex does not activate a reaction involving the effector itself: it is assumed only to bring about a discrete reversible alteration of the molecular structure of the protein or *allosteric transition*, which modifies the properties of the active site, changing one or several of the kinetic parameters which characterize the biological activity of the protein.[145]

Two years after the appearance of this important paper, there appeared another, equally important, article by Monod, Wyman, and Changeux in which "a plausible" model was presented for the nature of allosteric transition. After distinguishing between "homotropic" effects (interactions between identical ligands) and "heterotropic" effects (interactions between different ligands), the general properties of allosteric proteins were defined as follows:

(1) Most allosteric systems are polymers, or rather oligomers, involving several identical units; (2) Allosteric interactions frequently appear to be correlated with alterations of the *quaternary* structure of the proteins (i.e. alterations of the bonding between subunits); (3) While heterotropic effects may be either positive or negative (i.e. co-operative or antagonistic), homotropic effects appear to be always co-operative; (4) Few, if any, allosteric systems exhibiting *only* heterotropic effects are known. In other words, co-operative homotropic effects are almost invariably observed with at least one of the two (or more) ligands of the system; (5) Conditions, or treatments, or mutations, which alter the heterotropic interactions also simultaneously alter the homotropic interactions.[146]

In the formulation of the mathematics based on this model, the kind of symmetry found for hemoglobin was assumed to apply to oligomeric enzymes, with two conformational states (denoted R and T),

145. Monod et al. (1963), p. 307. In this article, the 1951 paper by Wyman and Allen was not cited. See also Debru (1983).

146. Monod et al. (1965), p. 89. An alternative model, involving sequential change, was proposed by Koshland et al. (1966). As Wyman (1972) has noted, both models are special cases of a more general theory of the cooperative binding of ligands.

and the effect of the interaction of a ligand (or ligands) in regions of the protein distant from the active site on the catalytic properties of such enzymes was expressed in terms of an equilibrium constant *(L)* for the R-T transition and equilibrium constants (K_R and K_T) for the binding of a ligand at the same site of the protein in each of the two conformational states. In oligomeric enzymes, such chemical equilibria are considered to be "linked" so that a ligand which binds preferentially to the T state will promote the transition from R to T. Wyman had previously developed the theory of linked equilibria on the basis of Gibbs's thermodynamics, and had applied the theory to the behavior of hemoglobin.[147] Recently, X-ray crystallographic studies on at least two oligomeric enzymes (muscle glycogen phosphorylase and aspartate transcarbamoylase) have provided structural data which have been interpreted in terms of this model.[148]

These recent achievements, and many other important experimental studies on the catalytic properties of oligomeric enzymes, attest to the stimulus provided by the Monod-Wyman-Changeux model to fruitful research. I believe, however, that it is misleading to state that this model "provides molecular biologists with a tool of the greatest power in their attempts to explain the biological in terms of the chemical." Apart from the question of the extent to which the model will be found to be generally applicable in more searching investigations of protein structure and function, it should be recalled that equilibrium thermodynamics can offer only a limited theoretical basis for the explanation of the chemistry of biological processes. It has long been recognized that, in a biological system (whether it be a bacterial cell or an animal or plant organism), the apparent "constancy of the internal environment" represents a "steady state" with the inflow and outflow of various forms of energy and chemical matter, and that the exchange of energy and chemical matter within parts of the biological system involves processes whose rates and oscillations require explanation in terms more complex than those used to describe the equilibria and their linkage in individual chemical components or reactions in that system. There can be no question of the importance, in biochemical thought and experimentation, of the thermodynamics developed by

147. Wyman (1948, 1964).

148. Schachman (1988); Barford and Johnson (1989); Kantrowitz and Lipscomb (1990).

Hermann Helmholtz and Josiah Willard Gibbs, but there is, I believe, also no doubt that few principles of mathematical physics have been so consistently misused in the writings of philosophically-minded biologists. I therefore introduce another brief digression, with special attention to the thermodynamic parameter of "entropy."

For example, in a book published in 1962, the distinguished microbiologist André Lwoff defined entropy as

> . . . a measure of atomic disorder. . . . The value of entropy is given by the formula: Entropy = $k \log D$, where k is the Boltzmann constant and D is the measure of the atomic disorder. The state of rest which an isolated system tends to reach corresponds to a maximum entropy, that is, to a maximum disorder. Now entropy, being a measure of disorder, is related to probability. In a closed system evolving from a less probable to a more probable state, the probability increases, and so does entropy. The relation between entropy and probability is given by the Boltzmann-Planck formula: $S = k \ln P$, where S is the entropy, k the Boltzmann constant expressed in ergs per degree centigrade, that is 1.38×10^{-16}, and P the number of "elementary complexions." "Elementary complexions" correspond to the "discrete configurations," that is, to the jumps of the atomic system from one metastable structure to another. . . . Physicists have found it convenient to have an expression for its reverse, that is degree of the atomic order or the availability for work. This negative entropy, "negentropy," represents the *quality* or *grade* of energy. And we may admit with Schrödinger that if D is a measure of atomic disorder, its reciprocal $1/D$ will be a measure of atomic order . . . The biologist currently speaks of information for the synthesis of a given enzyme. The concept of probability has disappeared, and the idea of quality, the specific value, is included in the biological concept of genetic information. The gene as a machine, or organization, contains information that is negative entropy. But, for the biologist, "genetic information" refers to a given actual structure or order of the hereditary material and not to the negative entropy of this structure. . . . The [biological] organism, according to Schrödinger, maintains its order by "sucking orderliness" from its environment. For the animal, the sources of orderliness are more-or-less complex or-

ganic compounds; for the plant, the source of orderliness is the sunlight. This type of reasoning has been endorsed by Brillouin, who writes, "If a living organism needs food, it is only for the negentropy which it can get from it and which is needed to make up for the losses due to mechanical work done, or simple degradation processes in the living system. Energy in the food does not really matter, since energy is conserved and never gets lost, but negentropy is the important matter." . . . The organism does not handle concepts of grade or logarithms of probabilities. The organism handles atoms or molecules and the energy of light or of chemical bonds. Nevertheless, some of the physicists I have consulted decided that Schrödinger's formula was perfectly acceptable, whereas others claimed that it did not make sense at all.[149]

Lwoff's definition of entropy, taken from Schrödinger's book *What is Life?*, is based on the principles of statistical mechanics and quantum mechanics, whose application has been of inestimable value in providing accurate data on the thermodynamic functions of simple molecules. As informed writers have emphasized, however, it is not appropriate to state that, as a general rule, the increase in entropy which occurs when an isolated system moves spontaneously toward equilibrium gives a measure of "disorder" or "randomness" in that system.[150] Moreover, serious criticism has been offered in regard to Brillouin's concept of "negentropy" and its interconvertibility with "information," as well as about the view that thermodynamic entropy is a measure of "ignorance."[151] I doubt whether many present-day experimenters in the biochemical sciences who use thermodynamics in their efforts to discern the chemical basis of the organization and regulation of biological structure and function have found much of value for their research in such high-flown rhetoric as "entropy is a measure of disorder" or the "isomorphism of entropy and information."

In view of the seeming adherence of some prominent biologists to such interpretations of the concept of entropy as that offered by Bril-

149. Lwoff (1962), pp. 89–97. This theme was developed in a more discursive fashion by Jacob (1970, 1973); for a critical review of Jacob's book, see Holmes (1977).

150. See, for example, McGlashan (1966) and Denbigh (1989a).

151. Brillouin (1962); Denbigh and Denbigh (1985).

louin, it may perhaps be necessary to call attention to the fact that, for modern working biochemists who learned Gibbs-Helmholtz thermodynamics from the book by Lewis and Randall,[152] entropy *(S)* is a function of the variables which define the state of a system, and that a measurable change in entropy (ΔS) equals Q/T, where Q is the heat absorbed during a reversible process and T is the absolute temperature (−273°C). For a system at constant temperature and pressure, this function is related to two other measurable thermodynamic functions: ΔG (the change in "free energy" available for "useful work") and ΔH (the change in heat content, or "enthalpy") by the equation $\Delta G = \Delta H - T\Delta S$. Moreover, for a chemical process in which a system is converted reversibly from state *A* to state *B*, the equilibrium constant *(K)* for that process is related to ΔG by the equation $\Delta G = -RT\ln K$ (where *R* is the so-called gas constant). To allow for comparisons between different systems of interest to a biochemist, the values for these three thermodynamic functions are given in terms of "standard states" (for a chemical reaction in aqueous solution at molal activities or, less precisely, of molar concentrations of all the reactants and products), usually designated as ΔG°, ΔH°, and ΔS°. Numerical values for each of these thermodynamic functions, as they apply to many individual reversible processes of biochemical interest, have been determined by means of various experimental methods.[153] When account was taken of the precision of such data, they have been of great value in the formulation of fruitful hypotheses, as for the sequence of electron-transfer reactions in the intracellular oxidation of metabolites such as succinate by molecular oxygen. In the search for chemical explanations for several other important biological processes, however, excessive reliance on thermodynamic reasoning has had, on occasion, a less salutary influence. At root, of course, is the inescapable fact that the rigorous theoretical framework provided by Gibbs and Helmholtz for the description of the differences in the energetic states of natural systems does not include any guide to the description of the process whereby such a change had been effected in a given system.

152. Lewis and Randall (1923).

153. See Krebs and Kornberg (1957) and Hinz (1986). There are many recent treatises on the application of equilibrium thermodynamics to the study of biochemical processes; I have found those of Edsall and Wyman (1958) and of Klotz (1986) to be especially useful.

On "Energy-Rich Phosphate Bonds"

During the mid-nineteenth century, the physiological speculations of Justus Liebig, based on the chemical knowledge of his time and on the measurements of the heat generated during animal respiration, encouraged the application of Germain Hess's "law of heat summation" to the thermochemistry of biological organisms. Moreover, the chemists Julius Thomsen and Marcelin Berthelot expounded the view that the evolution of heat (what we now denote as the change in enthalpy) which accompanies a chemical reaction is an accurate expression of the "affinity" of the components in that reaction. This view was soon challenged by Clausius's formulation of the second law of thermodynamics, and by van't Hoff 's definition of chemical affinity and of chemical equilibrium in terms of that law.[154] For the physiologists who were stimulated by Liebig's speculations, however, it was the "thermochemistry" of Thomsen and Berthelot which formed the theoretical basis for their calorimetric studies on animal metabolism. It was this approach, which combined measurement of the heat of combustion of each type of nutrient (protein, carbohydrate, fat) and of the respiratory exchange, as well as balance studies on the relationship between the dietary intake and excretory ouput, which guided the nutritional research of a series of noted German and American physiologists well into the twentieth century.[155] For others, notably Claude Bernard and Felix Hoppe-Seyler, this line of research had little value because it gave no information about the physiological or chemical processes within a biological organism.

The issue came to the fore in the study of muscular contraction, especially after the work of Fletcher and Hopkins, reported in 1907. Soon afterward, Archibald Vivian Hill felt obliged to note that

> unfortunately there have been, even among physiologists, many and grievous misconceptions in the application of the laws of thermodynamics; these have been due partly to the desire to make over-hastily a complete picture of the muscle machine, partly to the completely erroneous belief that the laws of thermodynamics apply only to heat engines and not to chemical en-

154. Lindauer (1962); Dolby (1984).
155. Lusk (1928); Holmes (1987a).

> gines, and that no information can be obtained from the Second Law as to the working of a chemical machine at uniform constant temperature. The muscle fibre has been treated as a heat engine when it is inconceivable that there are finite differences of temperature in it. . . . The muscle is undoubtedly a chemical machine working at constant temperature.[156]

This statement pointed, however, to a serious experimental problem. Whereas it was readily possible, by the use of the calorimeters then available, to estimate the enthalpy change (ΔH) in the conversion of glucose to lactate during muscular contraction, the estimate of the free energy change (ΔG) was more difficult, except in the case of reactions whose equilibria could be determined potentiometrically and required the application of Nernst's "third law" to the estimation of the entropy change. It is understandable, therefore, that Otto Meyerhof, who, with Hill, was until the early 1930s the leading investigator in the study of the energetics of muscular contraction, found support for his reliance on ΔH values in the report of Baron and Polanyi that at 310°K there is only a 13 percent difference between the heat of combustion of glucose (673,000 cal.) and the free energy of glucose (763,000 cal.).[157]

After the discovery in about 1930 of the role in muscular contraction of the acid-labile compounds creatine phosphate and adenosine triphosphate (ATP), Meyerhof determined the ΔH values for the hydrolysis of these and other phosphoryl compounds. He assumed that the $T\Delta S$ terms are relatively small, and concluded that the ΔG (pH 7) for the hydrolytic release of inorganic phosphate from the labile compounds is near −12,000 cal. per mole. During the 1930s, such estimates became important in the interpretation of new discoveries made in the study of the anaerobic breakdown of sugars (glycolysis) and in the study of utilization of the energy made available by oxidation processes for biochemical synthesis. A central concept was the idea of "energetic coupling" whereby a metabolic reaction (such as the cleavage of ATP to ADP + phosphate) with a relatively high negative ΔG of hydrolysis (a "strongly exergonic" process) is coupled to a reaction such as the phosphorylation of glu-

156. Hill (1912), p. 507.
157. Baron and Polanyi (1913); Meyerhof (1926), pp. 247–248.

cose to form glucose-6-phosphate, a compound whose ΔG of hydrolysis had been estimated to be about $-2{,}000$ cal. per mole.[158]

The concept of energetic coupling had been discussed long before by Wilhelm Ostwald. In 1900 he made a distinction between the transfer of chemical energy and the transfer of heat or electrical energy:

> A direct reciprocal transformation of chemical energies is only possible to the extent that chemical energies can be set in connection with each other, that is, within such processes that are represented by a stoichiometric equation. Coupled reactions of this kind may be distinguished from those which proceed independently of each other; their characteristic lies in the fact that they can be represented by a single chemical equation with definite integral coefficients. On the other hand, chemical processes that are not coupled cannot transfer energy to each other.[159]

However, when Meyerhof wrote in 1930 about his measurements of the relationship between oxygen uptake and glycogen resynthesis in muscle, he stated that "oxidation and resynthesis do not represent a chemically-coupled process, for which one can give a stoichiometric equation, but an energetically-coupled one."[160] Four years later, in the light of newer experimental knowledge, Jacob Parnas stated that

> the resynthesis of phosphocreatine and adenosine triphosphate is not linked to glycolysis as a whole, but to definite partial processes; and this leads further to the conclusion that this resynthesis does not involve a relationship which might be termed "energetic coupling," but more probably involves a transfer of phosphate residues from molecule to molecule.[161]

This concept of "group transfer" soon became an increasingly prominent feature of biochemical thought. In particular, the influential review article by Fritz Lipmann, published in 1941, introduced many biologists to the thermodynamics of group-transfer reactions,

158. Kalckar (1941).
159. Ostwald (1900), p. 250.
160. Meyerhof (1930), p. 38.
161. Parnas et al. (1934), p. 68.

with particular emphasis on the importance of ATP in "activating" chemical groups for such reactions.[162] He also extended Meyerhof's approach by developing the idea of the "energy-rich phosphate bond," which Lipmann denoted by means of a "squiggle" (ATP was written as A-P~P~P), as distinguished from the "energy-poor" bonds in esters (RCO-OR′) or amides (RCO-NHR′), to which he assigned ΔG values of hydrolysis of about 3,000 cal. per mole. Thus, he proposed that the biochemical function of ATP in the synthesis of the peptide bonds of proteins is to convert the carboxyl groups of amino acid units into reactive anhydrides (RCO-OP), which then undergo polymerization. As Lipmann expressed it some years later,

> Phosphate bonding is used . . . in the metabolic machinery as a means to parcel and transfer energy. Available energy is processed initially into energy-rich phosphate bonds of approximately 15,000 calories each and distributed through the adenylic-adenyl pyrophosphate and probably other systems. A key, or at least one of the keys, to its very general applicability appears to be the acid anhydride nature of the energy-rich phosphate bond. This makes it possible for the ATP to act as a common reagent for all these condensations involving the removal of water. In those reactions involving the carboxyl groups, this general statement can be qualified more precisely. It appears that here the primary reaction is the formation of an acid anhydride between carboxyl and phosphoric acid.[163]

Lipmann's distinction between "energy-rich" and "energy-poor" bonds and his emphasis on "phosphate bond energy" were questioned in succeeding years, as more reliable thermodynamic data became available.[164] Also, as the complex intracellular machinery for the synthesis of proteins began to be elucidated during the 1960s, it became evident that this process involves much more than the "activation" of carboxyl groups, and the alignment of activated amino acid units on a complementary "template." Although ATP plays a major role in protein synthesis, a large number of specific enzymes, not all of which have been adequately characterized, act in coordi-

162. Lipmann (1941).

163. Lipmann (1950), pp. 99–100.

164. See, for example, George and Rutman (1960); Carpenter (1960); Crabtree and Taylor (1979), pp. 369–371.

nated fashion as parts of intracellular organelles. Indeed, the chemical step in the formation of amide bonds in the RNA-directed assembly of peptide chains is the enzyme-catalyzed transfer of an amino acid unit from an ester to the amino group of the growing chain in a "transpeptidation" reaction.[165] Moreover, the applicability to intracellular systems of thermodynamic data for the enzyme-catalyzed hydrolysis of peptide bonds in aqueous solution came into question with the demonstration by Michael Laskowski, Jr., and his associates that in solvents of lower dielectric constant (for example, water-glycerol mixtures), the value of the equilibrium constant is shifted in the direction of peptide bond synthesis.[166]

Another aspect of thermodynamic theory, frequently misused by biochemists in the past, is the concept of "irreversibility." This word has had a rather different meaning in chemistry from that in evolutionary theory.[167] In biochemistry, the term has been applied to chemical reactions in which the equilibrium constant K is far in the direction of the products or, what is an equivalent statement, the value of ΔG is a relatively large negative number. Thus, the enzymatic hydrolysis of ATP to ADP and phosphate, with a ΔG (pH 7) of about −7,000 cal. per mole, was for a time considered to be an irreversible reaction. It has long been accepted that enzymes only accelerate the rate at which chemical equilibria are attained in isolated systems, and do not alter the numerical value of K. For example, the strongly exergonic hydrolysis of ATP by the "ATP-ase" of actomyosin filaments of mammalian muscle has been estimated to provide about 5,700 cal. of muscular work per mole of the ATP which had been hydrolyzed. This hydrolysis is "irreversible" only in the sense that the regeneration of ATP from ADP and phosphate (a strongly endergonic reaction) must be coupled to even "stronger" exergonic processes, such as those involved in oxidation of metabolites ("oxidative phosphorylation"). This coupled synthetic process, performed in the mitochondria of many organisms, has been the object of intensive study, especially since about 1950. Among the many important contributions was the demonstration by Efraim Racker that the catalytic agent in the condensation of ADP and phosphate is an ATP-ase whose action is coupled to the generation of hy-

165. Fruton (1988b).
166. Homandberg et al. (1978) and Fruton (1982b).
167. Denbigh (1989b).

drogen ions by the integrated and sequential operation of the previously known mitochondrial components involved in the transfer of electrons from metabolites to oxygen.[168]

These considerations are not unrelated to the widely held view among biochemists during the 1950s and 1960s that the intracellular enzymes whose function appeared to be the catalysis of biosynthetic reactions are of greater importance in the chemistry of life than the enzymes known to effect the breakdown of cell constituents. As it turned out, however, conclusions about the presumed intracellular function of isolated enzymes, based on data for the equilibrium constants of the individual reactions they catalyze, have repeatedly required revision. Such revision came in part from the discovery of previously unknown "activated" forms of the chemical units in particular biosynthetic reactions, and of new enzymes (or assemblies of enzymes) which catalyze such reactions more effectively than those previously thought to be the principal agents in these processes.[169] Moreover, intracellular proteolytic enzymes previously thought to be primarily involved in "death and decay" have been found to perform important functions in genetic regulation, viral replication, the transport of proteins across cell membranes, and the intracellular cleavage of proteins to generate peptide hormones.[170]

The Dynamics of Biochemical Processes

I will return later in this chapter to the question of the function of individual molecular constituents of living organisms, in relation to the idea of biological "purpose." More needs to said, however, about those features of biological organization whose explanation has involved, in addition to the application of the thermodynamics of chemical equilibria in isolated systems, the study of the rates and chemical mechanisms in the dynamics of processes which involve the catalytic activity of enzymes. Since the 1950s, notable advances

168. For a relatively recent summary, see Senior (1988). See also Racker (1976); Mitchell (1979); Walker et al. (1990).

169. Examples are the discoveries by Luis Leloir (1971) of uridine diphosphate glucose (UDPG) and its role in the biosynthesis of the polyglucosyl chains of glycogen, and by Arthur Kornberg (1988) that in *Escherichia coli* the synthesis of DNA involves the action of an assembly (DNA polymerase III) of ten distinctive subunits.

170. See Reich et al. (1975); Pontremoli and Melloni (1986); Rivett (1989); Kreil (1990).

have been made in this area of inquiry and have raised challenging problems for future investigation. The "lock and key" analogy suggested by Emil Fischer in 1894 for the binding of a substrate molecule in a suitably shaped cavity of an enzyme, and the concept of the enzyme-substrate complex developed by Victor Henri in 1902 and by Leonor Michaelis in 1913, had long guided the thought of enzyme chemists, but these ideas were extensively modified by the introduction of newer theoretical concepts and by new empirical discovery. Among the changes in theory was the recognition that, although enzymes merely catalyze the attainment of equilibrium in a reversible chemical reaction, the sequence of steps in the catalytic process constitutes a "steady-state" system in which the rates of the individual steps play a significant role. The introduction of the thermodynamic concept of the "transition state" into chemical kinetics, notably by Henry Eyring, emphasized the fact that, in some manner, an enzyme decreases the free energy of "activation" in the transformation of the initial enzyme-substrate complex into the reactive form, which is then converted into the products of the reaction. The development of what came to be called "physical-organic chemistry," especially by Christopher Ingold and Louis Hammett, brought to the fore discrete mechanisms (for example, acid-base catalysis) in organic chemical reactions, and such catalytic mechanisms were later found to be operative in enzyme-catalyzed reactions.[171]

With the parallel discovery and isolation of new enzymes and of previously unknown substrates (among which were derivatives of substances identified as "vitamins"), these theoretical concepts assumed even greater importance when it became possible, through the determination of the molecular structure of several enzyme proteins, to define the specific interactions between the active site groups of an enzyme and its substrates. These advances have encouraged the search for explanations of the ability of enzymes to increase the rates of otherwise sluggish chemical reactions. Particular attention has been paid to the relationships between the apparent equilibrium binding constant for the initial enzyme-substrate com-

171. See Bruice (1976); Walsh (1979); Westheimer (1985); Kraut (1988). Linus Pauling (1948, p. 709) briefly discussed the idea that the efficiency of enzymatic catalysis requires tight binding in the transition state. This idea was later fruitfully developed by Richard Wolfenden (1976), among others, through the demonstration that stable compounds whose structure resembles that of the presumed unstable transition-state form of the substrate are potent specific inhibitors of enzyme action.

plex and the catalytic rate. Although experimental data for some enzymes (for example, chymotrypsin) gave support to the widely held view that a lower value of K_M meant better catalysis,[172] the kinetic behavior of many other enzymes did not accord with this expectation. For example, pepsin (or papain) acts at very different rates in the hydrolysis of the same peptide bond in a series of substrates which differ in structure at a distance from the sensitive dipeptidyl unit, but for which the apparent binding energy is approximately the same.[173] These findings suggested that both the substrate and the active site in the initial enzyme-substrate complex undergo changes in conformation as a consequence of multiple non-covalent interactions which are weak individually, but which reinforce each other in the altered complex. Evidence for such changes came from the use of such techniques as fluorescence spectroscopy and of instruments to measure the "pre-steady-state kinetics" of the catalytic process, as well as from the X-ray crystallographic determination of changes in the relative positions of the amino acid side-chain groups involved in the catalytic mechanism when a substrate analogue is bound in the active site region of an enzyme.

More recently, the application of the technique of "site-directed mutagenesis," which makes possible structural changes in the active site region of some enzymes, has provided further evidence for the importance of the flexibility of such sites in enzymatic catalysis.[174] Moreover, the study of various protein systems has shown that, depending on the structure of the proteins and the ligands, "the kinds of flexibility are as diverse and complex as the chemical reactions themselves."[175] The thermodynamic interpretation of these phenomena strongly suggests that a portion of the actual binding energy (as distinct from the apparent binding energy) is used to make an important entropic contribution to the lowering of the energy state of the transition-state complex, and thus to the overall catalytic process.[176] Consequently, binding specificity and kinetic specificity in the action of enzymes appear to be, at the present state of our knowl-

172. Knowles (1965).

173. Fruton (1977).

174. Fersht (1988); see also Blow et al. (1986) and Knowles (1991). Schimmel (1990) has discussed some of uncertainties in the use of DNA technology for such studies.

175. Huber (1988), p. 88.

176. Jencks (1975), Page (1979).

edge, inextricably linked, and the mode of enzyme-substrate interaction becomes itself an important factor in the control of the catalytic rate. The relatively large size and compact structure of enzyme proteins, as compared with their catalytic sites, thus make for both stability of the whole and mobility of some of its parts. Clearly, similar considerations apply to the conformational changes in the quaternary structure of oligomeric enzymes induced by "allosteric" effectors, or of organized assemblies of catalytic proteins bearing active sites which differ in specificity. The complex kinetic behavior of such organized enzyme systems is likely to reflect more nearly that observed for intracellular chemical processes, for which non-equilibrium conditions prevail, than the behavior predicted on the basis of models based on equilibrium thermodynamic theory.[177] In further comment about Alexander Rosenberg's adulation of the theory of "allostery," it is therefore fair to say that the complexity of the dynamics of chemical processes in living systems still presents a major challenge in the "explanation of the biological in terms of the chemical."

A different challenge presented by the new knowledge about enzymes was anticipated in 1902 by Emil Fischer. Upon receiving a Nobel award for his syntheses of purines and sugars, he stated: "I foresee the day when physiological chemistry will not only make extensive use of the natural ferments, but it also will prepare artificial ferments for its purposes."[178] At that time, Fischer considered enzymes to be proteins, but he also believed that the molecular weight of proteins did not exceed about 5,000. During the course of his syntheses of polypeptides, he wrote to his former teacher, Adolf Baeyer: "My entire yearning is directed toward the first synthetic enzyme. If its preparation falls into my lap with the synthesis of a natural protein material, I will consider my mission fulfilled."[179] By 1910 Fischer had abandoned this effort, and eighty years later, when the methods of polypeptide synthesis far surpass the one Fischer had developed, the stepwise assembly of even a small enzyme, such as ribonuclease A (124 amino acid units), is a formidable task, even with modern automated instruments. Indeed, many enzyme chem-

177. See Peacocke (1983) for an account of efforts to develop theoretical models of kinetic self-organization in macromolecules and in biological systems.

178. Fischer (1924), p. 746.

179. Letter from Emil Fischer to Adolf Baeyer, 5 December 1905. This letter is in the Fischer collection, Bancroft Library, Berkeley, California.

ists have turned to the new DNA recombinant techniques employed in genetic engineering to synthesize variants ("mutants") of an enzyme under study.

Another promising development in the synthesis of new enzymes has come from the imaginative fusion of biological knowledge about the mode of the formation of specific antibodies with chemical knowledge of their structure and of organic-chemical mechanisms. By the use, as antigens, of stable "transition-state analogues" (see note 171), Richard Lerner, Peter Schultz, and Stephen Benkovic have succeeded in eliciting the formation of catalytic proteins of considerable efficiency and specificity.[180] Also, since about 1950, there have systematic efforts by some organic chemists to develop synthetic catalysts by attaching to macrocyclic compounds (cyclodextrins, polyethers) side chain groups of amino acid units which had been identified as functional parts of the active sites of enzymes (for example, the imidazolyl group of histidine). Such analogues of natural enzymes effected considerable rate acceleration in several types of biochemical reactions, such as the hydrolysis of ester bonds. This ambitious approach may lead to the construction of non-peptide frameworks whose three-dimensional structure and catalytic activity will mimic more closely those of some natural enzymes, and even to the synthesis of highly efficient new unnatural "enzymes."[181]

Some philosophers and philosophically-minded biologists and physicists may consider such chemical "tinkering" with proteins and enzymes to be, at best, a form of academic play or, at worst, dangerous tampering with what they conceive to be the "natural order." Such opinions should be evaluated in light of the history of the impact of synthetic organic chemistry on the medical sciences since the last quarter of the nineteenth century, and of the continuity of the study of "structure-function relationships." If, more recently, chemists have used as reagents specific enzymes in such operations as the conversion of pig insulin to human insulin, or recombinant DNA technology to effect what is now termed "site-directed mutagenesis," the objectives are, in principle, similar to the older efforts to synthesize new organic compounds of possible biological or medi-

180. Lerner and Benkovic (1988); Lerner et al. (1991).

181. For a recent account of efforts to synthesize artificial enzymes, see Lehn (1988); see also Mutter (1985).

cal interest, both for experimental purposes and as articles of commerce.[182]

Moreover, since about 1965, the euphoria generated by the explanation of the transmission of hereditary characters in terms of base-pairing of strands of DNA, and by the formulation of a "universal code" (now more modestly termed the "standard code") for the expression of genetic "information" in the biosynthesis of proteins, has given way to a more sober appreciation of the chemical complexity of biological systems. Through the fruitful experimental efforts of a vast number of research groups in the biochemical sciences, the accumulation of new empirical knowledge has brought many surprises to those who agreed with this statement made by Francis Crick in 1966:

> The ultimate aim of the modern movement in biology is in fact to explain *all* biology in terms of physics and chemistry. There is a very good reason for this. Since the revolution in physics in the mid-twenties, we have had a sound theoretical basis for chemistry and the relevant parts of physics. This is not to be so presumptuous as to say that our knowledge is absolutely complete. Nevertheless, quantum mechanics, together with our empirical knowledge of chemistry, appears to provide us with a "foundation of certainty" on which to build biology. . . . It is my argument that our present general knowledge of physics and chemistry is sufficient to act as an exceedingly solid foundation though, let me add, much of the *detailed* chemistry is incomplete and needs much further study.[183]

I have already mentioned some of the newer items of the "detailed chemistry" of proteins and enzymes, and have stressed the challenges in the study of the dynamics of their conformational change and catalytic action. In the field of nucleic acid biochemistry, similar problems have been posed by the demonstration that the three-dimensional structure of DNA exhibits much more variability and flexibility than was implicit in the Watson-Crick model for the fibrous form subjected to X-ray diffraction analysis.[184] The replication

182. Vane and Cuatrecasas (1984); Wright (1985); Anthony-Cahill et al. (1989).

183. Crick (1966), pp. 10–11.

184. Wells (1984); Häner and Dervan (1990); Holliday (1990). See also Stokes (1982).

of DNA strands involves a much more complex assembly of proteins than was indicated by the discovery in 1956 of the DNA polymerase I of *Escherichia coli*.[185] The enzymes which catalyze the synthesis of a DNA complementary to an RNA ("reverse transcriptases") are not, as first thought, limited to certain viruses, but are present in eukaryotes and play a role in the elongation of DNA chains.[186] The previously neglected enzymes (protein kinases) which catalyze the phosphorylation of the hydroxyl groups of serine, threonine, and tyrosine units, and the enzymes (protein phosphatases) which effect the hydrolysis of such phosphate esters, have been shown to be significant participants in many intracellular regulatory mechanisms and in the transmission of growth control signals.[187] The most notable of the recent "surprises" has been the discovery by Sidney Altman's research group that the bacterial enzyme (ribonuclease P) which catalyzes the specific conversion of precursors of transfer RNAs into their functional forms is itself an RNA.[188] This discovery not only provided a chemical explanation of the important observations of Thomas Cech's group on the "self-splicing" in the RNA of the protozoan *Tetrahymena thermopila*, but also presented to organic chemists a challenge to elucidate the mechanism of the action of such "RNA enzymes."[189] It may be expected that, in such studies, organic chemical synthesis will play a large role, as it did during the 1960s in the work of Gobind Khorana's group in establishing the "genetic code" through the synthesis of oligodeoxyribonucleotides of defined sequence, and as it has in the recent invention of methods for the automated synthesis of such compounds for use in recombinant DNA technology. Moreover, we now know that some biological organisms have enzyme systems which can effect the sequential assembly of amino acid units to form polypeptide chains without the participation of RNA.[190]

On Biochemical Function and Purpose

In hailing the recent achievements in the study of proteins, enzymes, and nucleic acids, and in noting some of the challenges these

185. Lehman and Kaguni (1989); see also note 169.
186. Shippen-Lentz and Blackburn (1990).
187. Hershey (1989); Cohen and Cohen (1989).
188. Guerrier-Takada et al. (1983); Altman (1990).
189. Zaug and Cech (1986); Westheimer (1986); Culp and Kitcher (1989).
190. Döhren and Kleinkauf (1988).

successes have presented, I must reiterate the obvious, namely that our ignorance of the "detailed chemistry" of biological organization far exceeds the seemingly enormous extent of our empirical knowledge. To affirm, in Edmund Wilson's words, that "to say *ignoramus* does not mean that we must also say *ignorabimus*" is, I believe, as valid today as it was in 1922. If, in the intervening years, philosophers and philosophically-minded biologists and physicists have shifted their allegiance from "neo-vitalism" or "organicism" to "anti-reductionism," and there is less talk about "bags of enzymes," these changes in attitude clearly reflect the impact on philosophical thought of the advances in the chemical explanation of important biological phenomena. In considering these advances to have been partial rather than "complete" chemical solutions of such problems as the transmission of hereditary characters or the intracellular synthesis and regulation of enzymes, one is not obliged, in my opinion, to accept the validity of such *ex cathedra* statements as "biochemistry could not be reduced to chemistry" (attributed to Karl Popper).[191] In the remainder of this chapter I will consider some aspects of the problem raised by statements of this kind.

First, I wish to comment about the view, expressed by the biochemist Steven Rose, that

> biochemists are different from organic and natural-product chemists in a number of important ways. *First*, for us the structure, sequence and molecular properties of substances derived from living organisms are not of great interest in their own right, but only insofar as they may be seen as providing information which casts light on the biological role of the substance. . . . *Second*, . . . we are likely to be less interested in the properties of 'pure' molecules in isolation, and more concerned with the ways in which they are involved in complex interactions with other molecules.[192]

This statement is part of Rose's argument against "philosophical reductionism" in the relationship between biochemistry (or molecular biology) and chemistry. I will return shortly to this issue, but feel obliged to note that the distinction Rose draws between biochemists and chemists is, I regret to say, arbitrary, superficial, and historically

191. Rose (1988), p. 160. See also Perutz (1986a).
192. Rose (1988), p. 161.

inaccurate. Many examples can be cited of bona fide biochemists who studied the "structure, sequence and molecular properties" of substances whose biological function was not understood, but which presented problems of biochemical significance. Among these biochemists were William Howard Stein and Stanford Moore, who determined the amino acid sequence of pancreatic ribonuclease A and established the nature of the catalytic groups in its active site, as well as Christian Anfinsen, whose work on the reversible inactivation of this enzyme laid the experimental groundwork for the study of the problem of protein folding. At the time these advances were made, the presumed "biological role" of ribonuclease A as an agent of the intestinal digestion of dietary RNA seemed to be superfluous, since it was known that all the components of nucleic acids are readily synthesized in animal organisms. The appreciation of the wider functional significance of the ribonucleases only came some years later.[193] The reason why Stein, Moore, and Anfinsen, along with Frederic Richards, Klaus Hofmann, and others, chose to study ribonuclease A was that it was the smallest known enzyme protein and was readily available in crystalline form. Similarly, although the hormonal nature of insulin was well known in 1945, it was not the biological role but the seemingly small size, crystalline nature, and availability of cattle insulin that determined the choice of this protein for Sanger's studies (described in the previous chapter) on its amino acid sequence. Moreover, there were bona fide organic chemists, notably Heinrich Wieland and Richard Willstätter, who concerned themselves actively with such problems as biological oxidations or photosynthesis. I cannot escape the impression that Rose's philosophical attitude may have been influenced by the kinds of institutional problems with which older biochemists are well acquainted. It is true that before the 1930s some leading organic chemists, particularly in England and Germany, took a dim view of biochemistry. It is also true, however, that the principal opposition to granting biochemists independent status in German universities came from senior members of the medical faculties who considered the subject to be merely an adjunct to physiology, pathology, hygiene, or internal medicine. Moreover, before the 1930s, the emphasis on the clinical and nutritional aspects of biochemistry in American medical schools, where there were many professors of

193. Benner and Allermann (1989).

physiological chemistry, was not favorable to the pursuit of the kind of biochemical research which had begun to flourish in Europe.[194]

Steven Rose's article about Karl Popper's pronouncement in 1986 that "biochemistry could not be reduced to chemistry" was followed by a small flurry of letters from biochemists, including one from Max Perutz, who reported that Popper had told him "that in his view a chemical reaction *in vivo* is different from the same reaction *in vitro*, because *in vivo* it works with a purpose, just as the combustion of petrol in a test tube is different from its combustion in the engine of a motor car."[195] There are, no doubt, many scientists (and perhaps philosophers) who may find Popper's statement to be unworthy of serious discussion at a time when the knowledge gained through the interplay of biological and chemical thought and practice had reached a stage of development far in advance of that available to Aristotle or to Immanuel Kant. As was cogently stated by Ernest Nagel,

> The question whether a given science is reducible to another cannot in the abstract be usefully raised without reference to some particular stage of development of the two disciplines. Questions about reducibility can be profitably discussed only if they are made definite by specifying the established content at a given date of the sciences under consideration.[196]

Since Popper's view bears a striking resemblance to the one offered by Aristotle, recent analysis of Aristotle's biological thought merits some attention. For example, Allan Gotthelf has concluded that

> Aristotle's teleology—his thesis that the development, structure, and functioning of a living organism are for the sake of something—is a central tenet of his thought. It is a corollary of the 'irreducible potential' interpretation of his conception of final causality that this thesis is *factual* or *empirical* in character: it is a conclusion drawn from observation of nature and not a premise brought to it. Philosophers of science today are in in-

194. Kohler (1982).
195. Perutz (1988).
196. Nagel (1961), pp. 361–362.

> creasing agreement that the question of reduction is an empirical one; they insist that one cannot legislate the precise form of the laws in which our understanding of nature is expressed. Aristotle's attitude is similar: he does not attempt to legislate a priori the particular form which a successful account of the natures and potentials of living organisms must take. . . . There is nothing in the fundamentals of Aristotle's philosophy, and nothing in his philosophical or scientific method, which would prohibit the adoption of a reducibility thesis, should the scientific evidence be judged to warrant it. But Aristotle did not believe that the evidence warranted it. On the contrary . . . he thought that a reduction of biology to chemistry could not be accomplished, and thus that organic development involves the actualization of an irreducible potential for form—which is to say, that it is a process *for the sake of* its end, irreducibility of potential for form being the core of Aristotle's conception of being *for the sake of something.*[197]

It is of interest here to note that the term *eidos*, which Aristotle used in various ways in his classification of animals,[198] and which has usually been defined as "form" or "species," was considered by Max Delbrück (and others) to be conceptually equivalent to the modern view of the genetic program which controls the development of the phenotype.[199]

In his *Critique of Teleological Judgement* (1790), Kant offered the biological principle that

> *an organized natural product is one in which every part is reciprocally both end and means.* In such a product nothing is in vain, without an end, or to be ascribed to a blind mechanism of nature. It is true that the occasion for adopting this principle must be derived from experience—from such experience, namely, as is methodically arranged and is called observation. But owing to the universality and necessity which that principle predicates of such finality, it cannot rest merely on empirical grounds, but must have some underlying *a priori*

197. Gotthelf (1987), pp. 229–230.
198. Pellegrin (1987).
199. Delbrück (1971); Grene (1976), pp. 19–20; Mayr (1982), p. 637.

> principle. . . . Indeed this conception leads reason into an order of things entirely different from that of a mere mechanism of nature, which *mere mechanism* no longer proves adequate in this domain. An idea has to underlie the possibility of a natural product. But this idea is an absolute unity of the representation, whereas the material is a plurality of things that of itself can afford no definite unity of composition. . . . It is no doubt the case that in an animal body, for example, many parts might be explained as accretions on simple mechanical laws (as skin, bone, hair). Yet the cause that accumulates the appropriate material, modifies and fashions it, and deposits it in its proper place, must always be estimated teleologically. Hence, everything in the body must be regarded as organized, and everything, also, in a certain relation to the thing is itself in turn an organ.[200]

In quoting this passage, I am aware of Kant's discussion of the contradiction ("antinomy") between the *idea* of the possibility of finding, through "experience," mechanistic explanations of the properties of biological organisms and the *idea* of the impossibility of finding such explanations. As I understand Kant's "resolution of the antinomy," it involves the conjecture that when certain biological phenomena cannot be explained mechanistically, teleological explanations are useful creations of the human mind. However, it would seem that, like many philosophers before and after him, Kant believed that some important properties of organized biological systems were not likely to be explained in non-teleological terms.

To put it over-simply, the debate over "means" (function) and "ends" (purpose) took different shape in Aristotle versus Democritos, Kant versus Descartes, and Popper versus Perutz, because at each successive stage the extent of reliable biological and chemical knowledge was much greater than before. In his valuable treatise, Ernest Nagel provided an impressive argument for the view that teleological explanations can be reduced to non-teleological explanations, and that the question whether biology can be reduced to chemistry is an empirical one.[201] Morton Beckner has drawn a clear distinction between purposive activity and functional activity, but

200. Kant (1928), pp. 24–26. For analyses of Kant's conception of biological methodology, see Zumbach (1984), McLaughlin (1989), and Körner (1990), pp. 207–214. Roll-Hansen (1976) has compared Claude Bernard's ideas with those of Kant.

201. Nagel (1961), pp. 398–446.

has also offered a modest defense of the use of teleological language in the description of biological function, a practice not uncommon among present-day molecular biologists.[202] Indeed, Ernst Mayr's view that there are now two distinct kinds of biology, functional biology and evolutionary biology, attests to his acceptance of the role of the interplay of chemistry and biology since 1800 in "explaining" many physiological processes which Aristotle considered to be "goal-seeking."[203] As in the past, "anti-reductionists" now call attention to our vast ignorance about cellular physiology, embryological development, brain function, population genetics, and other areas of biology as evidence for the "irreducibility" of these phenomena to chemistry. It would be idle to speculate whether further systematic biochemical research, with the aid of new chemical methods and physical instruments as well as hitherto unexamined living organisms, will provide reliable new empirical knowledge which will open new fruitful lines of investigation of these biological phenomena. However, no matter how significant the results of such research may turn out to be in shifting the focus of philosophical attention, it is likely that there will continue to appear, as in the recent past, writings heavily laden with anthropocentric, and even theological, bias and which assert "purpose" (in the sense of quasi-mental or divine design) to be an essential attribute of the biological world. It may also be expected that the debate about the role of "epistemological reductionism" in the interplay of biological and chemical thought and practice will continue to provide occasion for future symposia, as well many books and journal articles.[204]

One recurrent feature of recent philosophical writings about the reduction (or irreducibility) of biology to physics and chemistry (or simply to physics) is the implication that, except for some details, chemistry has already been largely "reduced" to physics.[205] I doubt

202. Beckner (1969, 1971). For a useful discussion of "purpose" and "function" in biological explanation, see Simon (1971), pp. 74–78.

203. Mayr (1961) and (1982), pp. 60–71. This distinction is implicit in the acceptance by most philosophical biologists of the validity of "ontological reductionism" and "methodological reductionism." At issue is the question of "epistemological reductionism"; see Ayala (1974), Hoyningen-Huene (1989).

204. Examples of the relatively recent contributions to the debate which, in my opinion, are the more valuable ones include those by Ayala (1968), Hull (1972, 1974), Schaffner (1974), Roll-Hansen (1978, 1979), Fuerst (1982), and Rosenberg (1989). A useful bibliography has been provided by Rosenberg (1985), pp. 266–271.

205. For examples of both sides of the debate, see Crick (1966), p. 10, and Ayala (1968), p. 208.

whether many organic chemists would accept the validity of this view. It is true that today quantum mechanics is an essential part of chemical theory, but it is also true that there have been at least two competing theoretical treatments of the covalent bond, the "valence-bond theory," associated with the names of Heitler, London, Slater, and Pauling, and the "molecular-orbital theory," associated with the names of Hund and Mulliken.[206] Moreover, much still remains to be learned about the physical nature of non-covalent bonding within and between organic molecules, a matter of obvious importance in the organization of structure and function in biological systems. Indeed, in its conceptual framework, most of present-day organic chemistry is no more a "reductionist" discipline than most of present-day biology. What is ignored in recent writings on the philosophy of biology is that, in the historical development of organic chemistry, the study of the relationship between the "whole" and the "parts" of natural substances involved not only theoretical speculation, but also the invention of methods to test organic-chemical theories by the artificial synthesis of the "whole" in a manner consistent with the existing, in large part metaphysical, conceptual framework of the discipline. In that respect, it may perhaps not be too far-fetched to compare the present state of the art of biotechnology to that of organic chemistry during the mid-nineteenth century.

Another feature of the recent writings by philosophers of biology about the issue of reductionism is their concentration on the development of molecular genetics since about 1950 and its relevance to evolutionary theory. Only occasionally does one find in a volume devoted to the philosophy of biology an appreciation of the historical continuity of the debate, as in the article by June Goodfield, who began by stating that

> I am overpowered by a feeling of *déjà vu* verging at times on the very edge of intellectual impotence. 'Reductionism'; 'antireductionism'; 'beyond reductionism'; 'holism.' We have seen these words and heard the accompanying arguments so many times before. The issue is a very old one, recurring in various forms with unfailing regularity throughout biological history, and the

206. See Coulson (1961), p. 71; Liegener and Del Re (1987); Klein and Trinajstic (1990).

feeling of impotence arises because, after all this time, the issue never seems to get any clearer.[207]

In her essay Goodfield called attention to the nineteenth-century arguments, some of which were mentioned earlier in this chapter.

In considering the problem of biochemical function, and the relationship of this concept to that of biological purpose, it is useful, I believe, to recall the history of the investigation of seemingly simple chemical processes as they occur in biological systems. One such process is the formation, in the mammalian liver, of urinary urea (NH_2-CO-NH_2) from ammonium ions (NH_4^+) derived from amino acids and respiratory carbon dioxide (CO_2). During the period 1860–1930, this process was studied by many physiologists and chemists. Various hypotheses were advanced, attacked, and defended, but all of them were recognized to be inadequate after Hans Krebs showed in 1932 that the process, as it occurs in surviving liver slices, is far more complex than had been previously imagined. Since then, the biosynthesis of larger chemical constituents of biological systems, such as cholesterol or the heme of hemoglobin, has been been found to be even more intricate than that of urea. Biochemists have accepted as a general principle the idea that the processes of natural selection have led, through millennia, to the elaboration of chemical processes whose existence could have been predicted only occasionally (and even then only partially) on the basis of the available knowledge of the chemical properties of the initial and final components. Biochemists have also learned that the chemical capacity of an isolated enzyme to catalyze the formation of polypeptides, polynucleotides, or polysaccharides does not necessarily establish that enzyme as the one responsible for the intracellular formation of such polymers. Although the reaction which is effected in each case is clearly evident from purely chemical studies, in most cases the enzymes now believed to be the actual catalysts in intracellular processes, or the intermediate substances in such processes, are different (and less simple in structure) from those initially assumed, solely on chemical grounds, to be involved in the physiological process under study. These changes in accepted belief are not consid-

207. Goodfield (1975), p. 65.

ered to be setbacks (except by some protagonists of a disproved hypothesis), but as advances in biochemical knowledge.

On Specificity and Individuality

The seeming lack of historical perspective in many of the recent writings on the philosophy of biology is, in my opinion, related to a readiness to accept rather uncritically the validity of the historical accounts and philosophical interpretations offered by some of the investigators (and their disciples) whose research played large roles in the development of the present conceptual framework of the biochemical sciences. A striking feature of that development was, and continues to be, the participation of scientists whose diverse educational backgrounds and previous research interests brought into the interplay of biological and chemical thought and practice different attitudes to some of the traditional general concepts which had long animated that area of scientific inquiry. One of these concepts is that of specificity, which has figured prominently in past controversy over the question whether human diseases are caused by "specific" entities or by disturbance of a "natural balance." During the nineteenth century, the ancient Paracelsian view that specific chemicals are required to cure specific diseases[208] acquired some respectability with the transformation of the practical art of pharmacy, through advances in both animal physiology and organic chemistry, into the biochemical science of pharmacology.[209] The question of the relationship between biological specificity, made evident by morphological and physiological discoveries, and the specific interactions of the chemical substances of which biological organisms were known to be composed was brought forward most prominently at the turn of the century in two separate areas of scientific inquiry: the mode of enzyme action and the immunological response to a toxic agent. I have already mentioned Emil Fischer's "lock-and-key" analogy to describe the specificity of enzymatic catalysis in terms of a complementary structural relationship between interacting chemical substances, and the subsequent modification and elaboration of this theory. Such "chemical specificity" is seldom absolute in the sense

208. Lindeboom (1957).
209. Bynum (1970); Parascandola (1974, 1975).

that a given enzyme will act to a measurable degree only on a single substrate. In an organized multi-enzyme system, however, the overall selectivity in the production of a unique product is likely to be greater than the "relative specificity" of the individual components of that system.

The lock-and-key analogy played a considerable role in the historical development of ideas about immunological specificity. In my opinion, this area of biochemical inquiry represents the prime example of the fruitful (and contentious) interplay, for nearly a hundred years, of the concepts of biological specificity and chemical specificity. After the 1890s, the discovery by Émile Roux and by Emil von Behring that the serum from animals infected by pathogenic bacteria contain "antitoxins" to the toxic agents present in the bacterial cultures assumed greater importance than the idea offered by Elie Metchnikoff in the previous decade that particular blood cells are responsible for the defense of an animal organism against pathogenic microbes.[210]

During the first decade of this century, the most widely discussed theory of immunity was that of Paul Ehrlich, who had proposed that living protoplasm elaborates receptors which interact specifically with chemical groups of toxins in a kind of lock-and-key process analogous to that postulated by Fischer for the action of enzymes. Ehrlich's view was opposed by the physical chemist Svante Arrhenius, who considered the toxin-antitoxin reaction to be an equilibrium process such as that in the interaction of weak acids and weak bases.[211] In response, Ehrlich wrote:

> In view of the extraordinary success which physical chemistry has scored, it is readily understood how tempting it was for so eminent a representative of this science as Arrhenius to apply its principles to the new field of immunity. I have always emphasized the chemical nature of the reaction, and am glad therefore that the attempt to apply these principles has been made. It has been demonstrated anew that the phenomena of animate nature represent merely the resultants of infinitely complex and variable actions, and that they differ herein from the exact sciences, whose problems can be treated mathematically. The for-

210. Metchnikoff (1901).
211. Wassermann (1910); Mazumdar (1974); Rubin (1980).

> mulas devised by Arrhenius and Madsen for the reaction of toxins and antitoxins explain absolutely nothing. . . . Neither do I believe that the phenomena observed in toxins and antitoxins bear any relation to the processes of colloid chemistry. . . . Structural chemistry, on the other hand, has not only served to explain all the phenomena in immunity studies, but has also proved a valuable guide in indicating the lines along which further progress might be made. The limitations of colloid chemistry have already manifested themselves, and enthusiastic advocates of this science have been compelled to assume the existence of specific atom groupings in accordance with my views. I therefore see no reason for abandoning the views expressed in my receptor theory, a theory in complete accord with the principles of synthetic chemistry.[212]

The specificity of the interaction *in vitro* of chemically modified protein "antigens" with the serum "antibodies" elicited upon the injection of such proteins into suitable animals was first described by Obermayer and Pick in 1906. This approach was notably extended during the 1920s by Karl Landsteiner, who brought to his immunological research experience in organic chemistry, gained through work (1891–1893) in the laboratories of Emil Fischer and Adolf Baeyer. In particular, Landsteiner showed that the antigenic specificity of a modified protein could be determined by the nature of the chemical substituent ("hapten").[213] Also, by the 1930s it was recognized that the specificity of the immune response was not limited to proteins or artificial haptens, but also encompassed carbohydrates and nucleic acids. This vast diversity in the chemical repertoire of immunological specificity, together with many unexplained biological features of the immune response, invited renewed speculation about the process of antibody formation.[214] Between 1910 and 1940 there was a succession of hypotheses based on the idea that specific antibodies are generated by a mechanism whereby the antigen directly modifies the structure of a normal serum protein to confer upon it the appropriate specificity. Among the investigators whose names are associated with various forms of this "template"

212. Ehrlich (1910), p. 578.
213. Landsteiner (1945).
214. See Silverstein (1982, 1989) and Nossal (1986).

theory were Oskar Bail, Friedrich Breinl, Felix Haurowitz, Stuart Mudd, William Topley, and Linus Pauling. It is of special interest to note that during the late 1930s Pauling (with encouragement from Landsteiner) formulated a version of the theory (based on the "rule of parsimony") which involved the assumption that

> the number of configurations accessible to the polypeptide chain is so great as to provide an explanation of the ability of an animal to form antibodies with considerable specificity for an apparently unlimited number of different antigens, without the necessity of invoking also a variation in the amino acid composition or amino-acid order.[215]

A prediction of this theory was that, in the presence of an antigen, a partially unfolded ("denatured") serum globulin would refold *in vitro* into a specific antibody. As in the case of previous claims to have effected the artificial formation of antibodies, the claim by Pauling and Dan Campbell that they had confirmed the prediction proved to be irreproducible.[216] Indeed, some years later, Edgar Haber showed that a partially denatured antibody fragment can regain its specificity *in vitro* in the absence of the appropriate antigen.[217] Clearly, Pauling's concept of the role of "molecular complementarity" in biological specificity had led him to an incorrect theory of antibody formation.

During the 1930s, one of the opponents of the organic-chemical approach to problems of immunity was Ludwik Fleck, a relatively obscure Polish serologist, who characterized the views of Ehrlich and his followers as "chemical delusions."[218] Fleck's book was rescued from oblivion in 1962, when Thomas Kuhn referred to it in the preface to *The Structure of Scientific Revolutions;* the book attracted considerable attention among sociologists and philosophers of science because Fleck had expounded views that many of

215. Pauling (1940), p. 2644.

216. Pauling and Campbell (1942). For earlier claims, see Manwaring (1930). See also Pauling (1970, 1974); in these retrospective accounts of his scientific achievements, Pauling wrote about his efforts in the field of immunochemistry, but omitted mention of the claim to have produced antibodies *in vitro*.

217. Haber (1964).

218. Fleck (1935, 1979). For recent discussions of Fleck's views, see Schnelle et al. (1983), Cohen and Toulmin (1986), Neumann (1989), and especially Belt and Gremmen (1990).

them considered to have been anticipations of opinions, current during the 1960s, about the social determinants in the growth of scientific knowledge. From a lengthy (and perhaps questionable) analysis of the development of the Wassermann test for syphilis, Fleck drew the generalization that the thought of individual scientists is subordinate to what he called the *Denkstil* of a *Denkkollektiv*. In my opinion, the recent enthusiastic discovery by sociologists of Fleck's ideas is a rather more interesting subject of future historical study than the negligible impact of his serological views of the 1930s on his professional colleagues. Fleck's career, as a Polish Jew who survived incarceration in Nazi concentration camps, invites not only sympathetic understanding but also critical biographical study of a scientist who sought to define important questions about the nature of scientific investigation.

Although the cellular theory of Metchnikoff and the protoplasmic ideas of Ehrlich about the formation of antibodies lingered in the background of immunological research for several decades, it was not until after 1945, partly in reaction to Pauling's theory, that a more biological approach once again became prominent. The leading figures in this development were Macfarlane Burnet, Niels Jerne, Peter Medawar, Joshua Lederberg, and David Talmage. In the theory which emerged from their writings during the 1950s, the emphasis was on antibody-producing cells and the role of an antigen in eliciting the multiplication of those cells which had the genetic capacity to produce an antibody specific for the antigen. This theory of "clonal selection" represents, in my view, as significant an advance in modern biological thought as the theory of the gene, for it raised new problems for the continued interplay of biology and chemistry. I cannot attempt to summarize the present state of research in this active field of investigation except to call attention to the biochemical contributions to its development. In particular, the determination of the arrangement of the polypeptide chains of the immunoglobulins, and of the amino acid sequences of these chains, principally by Rodney Porter and Gerald Edelman, attest to the power of the methodology introduced into protein chemistry by Sanger and later developed by Stein and Moore. It is now known that an immunoglobulin molecule is composed of four peptide chains arranged in a Y-like structure; each of the chains has an amino-terminal peptide segment which is variable in amino-acid composition, to form two specific antigen-binding sites. Moreover, the invention of new techniques has opened new areas of investigation of the immune response by both chemical and genetic methods. Of special importance has been the method, developed by Cesar Milstein and Georges Köhler, in which myeloma cells (each of which produces a homogeneous antibody) are fused artificially to spleen cells from an immunized animal to form "hybridoma cells."[219] The present conceptual framework of immunology thus embodies elements of both of the seemingly disparate theories advanced by Metchnikoff and Ehrlich. The term "epitope," introduced by Niels Jerne, refers to the portion of the antigen which confers specificity on the manner in which it elicits the production of an antibody, and hence is related to earlier terms such as "hapten," "immunogenic determinant," or "antigenic pattern."

The recent development of immunology provides additional evidence for the view that the study of biological specificity can be illuminated by the concurrent study of the intimate structure and specific interactions of chemical molecules. Can such fruitful interplay of biological and chemical thought and experiment also be expected in the field of embryology, which, as I indicated earlier, yielded primacy to genetics? Although some hope is offered by recent advances in the study of the genetic control of embryonic development,[220] the problem of the specificity of the processes in the transformation of a fertilized egg into an adult organism still presents formidable challenges. It is appropriate at this point to insert the opinion expressed during the 1940s by the noted embryologist Paul Weiss about the biologist's conception of "specificity":

> In its common connotation, the term refers to that relation between two systems which enables members of one system to exert a *discriminative* effect upon certain members only of the other; it implies *selectivity* of action and reaction even in the absence of separate channels from the acting to the reacting members. A chemical that bathes all tissues, but affects some of them with disproportionately greater potency than others, will be considered as specific for the affected tissue in that sense. By definition, selectivity is the faculty of a process or a substance to activate, to alter the state of, or to combine with,

219. For a relatively recent summary, see Nisinoff (1985).
220. Gehring (1985).

> certain elements in preference to, and to the exclusion of, other elements of the same system. The basic criterion of selectivity, therefore, is the correspondence and mutual fitting between two properties. Primarily, the term specificity applies to this *correspondence*, and to neither of the interacting systems as such. By custom, however, it has acquired a secondary meaning signifying those properties of each system which make selectivity of interaction possible.[221]

In his experimental work, Weiss sought to identify the specific processes in the organization and differentiation of cells during embryonic development, but I believe it fair to say that his chief contribution was in defining the problems of embryology within the limits of the biochemical knowledge of his time.

Another statement about biological specificity which merits recollection is that of Jacques Monod, in 1947:

> One of the most characteristic tendencies of the present period in the development of biology, may perhaps be seen in the focussing of attention on problems of *specificity*. As emphasized by Weiss, in spite of the widely different connotations of the word, any kind of specificity must, of necessity, be associated with a particular, "specific," pattern in space or time or both. Thus, it is generally recognized that one of the main problems of modern biology is the understanding of the physical basis of specificity, and of the mechanisms by which specific molecular configurations (or multi-molecular patterns) are developed, maintained, and differentiated. The means, the experimental tools for this study, are found in those experiments which result in inducing the formation, or suppressing the synthesis, or modifying the distribution of a specific substance or substances.[222]

During the 1950s and 1960s, the term "biological specificity" seemed to have lost favor among some of the leading spokesmen for the new molecular biology. It appeared, however, in the remarkable book by Horace Judson, who based his account of what he termed "the revolution in biology" on conversations with many scientists

221. Weiss (1947), p. 236. See also Haraway (1976), pp. 147–187.
222. Monod (1947), p. 224.

and on thoughtful examination of the pertinent literature.[223] According to Judson,

> Biological specificity in its present sense had a number of forerunners. Today, three considerations subsume all the rest. The first is linearity: the long-chain biological molecules, both proteins and nucleic acids, are specific in sequence. The second is structure: the large biological molecules owe their functional specificity to specific configurations. . . . The third consideration is that linear specificity determines three-dimensional specificity.[224]

In his review of Judson's book, Gunther Stent expressed the opinion that

> as for "specificity," it remains, across disciplines, as vacuous a concept today as it was in the thirties. . . . In present-day immunology, for instance, the "specificity" of an antibody molecule is still viewed in terms of its physico-chemical affinity for an antigen, just as it was in the 1940s when Pauling proposed his theory of antibody formation. To this affinity the "specific" amino-acid sequence of the antibody bears a complicated and as yet unfathomed relation.[225]

Apart from the question of the accuracy of Stent's description of the state of immunological thought and empirical knowledge in 1979, or the rejection by others of the idea of molecular specificity in the study of embryonic development,[226] there can be little doubt that the age-old issues of vitalism versus mechanism, or reductionism versus holism or teleology, are still prominent in the interplay of the biological and chemical sciences. Indeed, the current use of such metaphors as "recognition," "fidelity," or "proofreading" in the description of the molecular interactions in some enzymatic pro-

223. Judson (1979). For three rather different assessments of this book, see Stent (1979), Edsall (1980), and Abir-Am (1985), pp. 94–101.

224. Judson (1979), p. 608.

225. Stent (1979), pp. 426–427.

226. See, for example, Sibatani (1981) and references cited therein.

cesses suggests that even "reductionist" biochemists are not entirely free of the taint of teleology.[227]

Whatever terms one may choose in place of "specificity" (as defined by Weiss), it may be expected that new chemical constituents of biological systems will be discovered and that new knowledge will be gained about the structure of and intimate interaction among the known and new substances.[228] As I suggested earlier, the greatest biochemical challenges still lie in the study of the mechanisms whereby the relative rates of the individual chemical reactions in the "wholes" and "parts" of living systems are specifically organized and regulated. Some biologists or philosophers may find a further increase in the (to them) staggering complexity of such empirical knowledge to be evidence of the futility of the biochemical enterprise, except perhaps in providing useful drugs for medical practice. It is likely that some theoretical physical chemists and computer experts will continue efforts to bring rational order into the ever-growing complexity through resort to model-building and systems analysis.[229] It would be idle to guess how theory and experiment in the biochemical sciences will develop in the future, especially since one cannot rule out the possibility that new (and perhaps accidental) empirical discovery may lead to the creation, in the laboratory, of an assembly of chemical substances which, though unlike any known microorganism, would be acknowledged by biologists to be "living." As was noted by Pirie, the past disproof of the "spontaneous generation" of living things from inert matter does not exclude this possibility.[230]

To concepts such as organization and specificity which have figured in the interplay of biological and chemical thought may be added that of "individuality." In previous sections of this chapter I have referred to the individual biological forms used in particular investigations. Among these forms were higher animals (including human subjects), plants such as the garden pea and maize, and

227. For an interesting discussion of this question, see Rosenberg (1985), pp. 255–265.

228. Among the surprising new discoveries is the demonstration that the "endothelium-derived relaxing factor" which mediates vasodilatation is nitric oxide, a toxic compound made enzymatically from arginine; see Bredt and Snyder (1990) and Butler (1990).

229. Berlinski (1976), pp. 156–178; Miller (1978); Peacocke (1983).

230. Pirie (1954).

organisms such as the vinegar fly, yeast, and the colon bacillus. Each of these organisms, or colonies of organisms, has been assigned a taxonomic name, based on a succession of classificatory schemes, and in recent years there have been at least three competing proposals.[231] I shall not attempt to summarize the history or present state of the methodology of biological systematics, except to note that there has been disagreement among evolutionary biologists and philosophers of biology concerning which entity should be considered the "individual" unit of natural selection (genotype, organism, species, population).[232] Moreover, it is not always clear what philosophical biologists mean by "individuality," since biological "laws" do not refer to specific individual organisms but to classes of such organisms. In connection with this uncertainty, it seems appropriate to call attention to the cogent statement by the philosopher Stephan Körner:

> The manner in which a person classifies the objects of his experience into highest classes or categories, the standards of intelligibility which he applies, and the metaphysical beliefs which he holds are intimately related. . . . Among the characteristically philosophical arguments some are used mainly in support of the categorial *status quo*, others mainly in the cause of categorial revolution, while still others are used opportunistically for either purpose.[233]

Apart from the important question of the relationship between the metaphysical beliefs which underlie "rational" classifications and the "natural" order of living things, it would seem that, for many biologists, the properties of biological organisms, and their interaction with one another and their environment, are inextricably linked with their "individuality."

In some recent writings, this attitude has been associated with "aesthetic recognition." For example, the philosopher of biology Marjorie Grene wrote:

231. Mayr (1968) and (1988), pp. 268–288. See also Bremer et al. (1990) for a complaint about the instability of current biological taxonomy.

232. See Rosenberg (1985), pp. 180–225; Ruse (1987); Cracraft (1989).

233. Körner (1970), pp. ix, 69.

> The practice of biology entails a recognition of pattern in a more pervasive sense than do physics and chemistry. Over and above the recognition of abstract patterns characteristic of the sciences of inanimate matter, the practice of biology demands the recognition of individual living things, and analysis in biology is always analysis *within* the context set by the existence of such individual living things.[234]

This affirmation raises several questions. If one grants that aesthetic recognition is common among naturalists, as well as common folk who find pleasure in zoos and aquariums, cannot a similar response be elicited in the mind of a geologist when he examines natural minerals, or in the mind of a non-biological scientist who contemplates the differences in the forms of hemoglobin crystals obtained from the blood of various animals?[235] Might the adulation accorded by some biologists to D'Arcy Thompson's treatment of the beautiful geometry of living things[236] also apply to the aesthetic recognition of symmetry in the constitution of chemical substances?[237]

It should be recalled that during the nineteenth century, with the great increase in the number of known "homogeneous" chemical substances, the problems of classification were prominent in chemical thought. The earlier ideas had been based, as in botany and zoology, on the qualitative similarity in the observable properties of such substances. It was not until the question of the "individuality" of a chemical material was attacked through the determination of the quantitative proportions of the elements in a compound that the problem of classification assumed a different form. The change came about as a consequence of the interplay of many theories and much experimental effort. Thus the discovery, through quantitative analysis, that two substances which differ in their properties can have the same elementary composition (they were termed "isomers") and the later explanation of qualitative differences between isomeric substances (such as their ability to rotate the plane of polarized light, or "optical activity") in terms of the "metaphysical" concepts of valence, structure, and stereochemistry laid the groundwork for

234. Grene (1961), p. 195; see also Pantin (1954).
235. See Reichert and Brown (1909) and Drabkin (1975).
236. Gould (1976).
237. Hargittai and Hargittai (1986).

a classification of carbon compounds. Of course, the basic contribution was the development of the periodic table of the elements, by Dmitri Mendeleev and Lothar Meyer and their precursors.[238] In short, the problem of the classification of chemical substances had largely been resolved by applying the term "individual" to the "species" to which it was assigned, whether the substance had been obtained in crystalline or amorphous form. Indeed, some mineralogists, with their primary interest in crystalline structure, have considered a crystal to be a "true individual."[239] However, as in biological classification, there have been difficulties presented by the existence of "mixed crystals," azeotropic mixtures, and the like. Regrettably, these aspects of the historical development of chemistry have received relatively little attention from philosophers of science. A notable exception was Gaston Bachelard, among whose writings is the following passage:

> We thus face a question whose philosophical importance is inescapable: how can the knowledge of a particular substance be refined, specified, multiplied by the knowledge of a different substance, or better still by the extensive knowledge of a group of substances? . . . One may say, in a paradoxical manner, that one reduces diversity by increasing it, because by introducing new substances into an incompletely known series of substances one substitutes the knowledge of the series for the knowledge of the individual substances.[240]

In offering these comments about the analogy of the problems of classification in chemistry to those in biology, I readily acknowledge the much greater difficulty of the question of the relationship between artificial biological classification, based on empirical knowledge of the anatomy, physiology (including biochemistry), embryology, genealogy, and phylogeny of the vast diversity of known biological organisms, and the classes considered since the time of Aristotle to be of a "natural" kind. I return to this question in the final chapter in this book, which deals with the languages of the biochemical sciences.

238. Cassebaum (1971); Cassebaum and Kauffman (1971).
239. Hooykaas (1952, 1958); Mauskopf (1976); Laudan (1989).
240. Bachelard (1973), pp. 23–24. See also Bachelard (1972).

Evolutionary Theory and the "Unity of Biology"

The modern interpretation of Darwin's theory of the mechanism of natural selection, together with new knowledge of the mechanisms of the transmission of hereditary characters, has provided the basis for the interpretation of data on the chemical diversification among individual members of a biological species. Thus, many of the differences among human beings with respect to their physiological characteristics and their susceptibility to disease have been described in terms of the detailed chemistry of protein enzymes and their interaction with each other and with other chemical constituents of the human body. Moreover, in a number of cases, the systematic study of the variations in the amino acid sequence of a particular protein which is present in a wide variety of organisms has provided data for the construction of a phylogenetic tree for these organisms and the formulation of theories about the evolutionary history of the gene locus associated with the synthesis of that protein.[241] An outstanding example of studies of this kind is the work of Clement Markert and his associates on the multiple molecular forms ("isozymes") of the enzyme lactate dehydrogenase, a tetrameric protein whose synthesis appears to involve a two-gene system.[242] More recently, the use of methods for DNA-RNA hybridization, for the determination of nucleotide sequences (as by the latest versions of the polymerase chain reaction), and for the specific enzymatic cleavage and ligation of internucleotide bonds has provided extensive data whose relevance to evolutionary theory has been widely discussed. One of the challenging questions is that presented by the finding that only a small section (perhaps 2 to 5 per cent) of mammalian DNA "codes" for identifiable proteins, and the role (or heritage) of the seemingly non-functional sections ("introns") has been a matter of active recent investigation.[243]

Other chemical aspects of evolutionary theory which have attracted considerable attention are the "prebiotic" evolution of organic chemical compounds, in particular nucleic acids and proteins; the "origin of life"; and the "self-organization" of chemical units into biologically functional units. During the 1920s Aleksandr

241. Sigman and Brazier (1980); see also Simpson (1964a).
242. Markert and Whitt (1968); Markert et al. (1976).
243. See, for example, Marchionni and Gilbert (1986).

Oparin and J. B. S. Haldane advocated the view that the oxygen-free ("reducing") atmosphere of the primitive earth favored the synthesis of simple organic compounds. Thirty years later much was made, therefore, of Stanley Miller's experiments, in which he subjected a mixture of methane, ammonia, hydrogen, and water to an electrical discharge (for a week), and his finding of some amino acids among the products. Also, by heating amino acids to 130°C, Sidney Fox obtained water-insoluble particles, which he named "proteinoids," and suggested that such bodies might have been precursors of living cells.[244] More recently, Christian de Duve has proposed a scheme of prebiotic chemical evolution, involving the prior formation of catalytic "multimers" which effect the formation of "energy-rich" thiol esters.[245] One does not have to be an old-fashioned organic chemist to view such speculations with skepticism. The problem of "self-organization" has been treated in elegant fashion by Manfred Eigen and others, and models have been proposed for the evolutionary conversion of the "primordial soup" into a system in which "informational" nucleic acids and proteins are assembled and selected in processes which involve both autocatalytic synthetic reactions and degradative reactions.[246] The relevance of such models, no matter how ingeniously they may have been contrived, to the actual history of biochemical evolution is uncertain, in view of the manifold possible chemical transformations of DNA and RNA (duplication, recombination, transposition, excision), and as long as "chance," "messiness," and "tinkering" remain prominent features of our view of the evolutionary process.[247]

To conclude this chapter, I offer some opinions about the past search for "unity" in the contemplation of the diversity of the increasing number (now about two million) of identified biological species. The idea of the "great chain of being," so brilliantly traced by Arthur Lovejoy,[248] was replaced during the nineteenth century by a series of non-theological concepts, the most important of which was Darwin's theory of natural selection. As noted earlier, there were also biologists who preferred to find the unity of biology in

244. Fox and Dose (1972); for a review, see Pirie (1974). See also Miller and Orgel (1974).

245. De Duve (1988).

246. Eigen and Schuster (1979).

247. Jacob (1982), pp. 27–46; Katz (1987).

248. Lovejoy (1936).

the cell-theory, with its attached concept of an "albuminoid protoplasm." Claude Bernard proclaimed that "there is only one kind of life, there is only one physiology for all living things."[249] Early in the twentieth century, the microbiologist Albert Jan Kluyver wrote of the "unity of biochemistry" in calling attention to the similarity of the then-known metabolic pathways in diverse biological organisms.[250] More recently, the geneticist Robert Hall Haynes stated: "The discovery of the chemical nature of the genetic material, and of the mechanisms of gene action, have provided biology with a unifying theoretical structure that complements, at the molecular level, that great overarching principle of unity in the world of life, natural selection."[251] And the paleontologist George Gaylord Simpson, in writing of the historical processes of biological evolution, stated:

> The results of these processes are systems different in kind from any nonliving systems and almost incomparably more complicated. They are not for that reason any less material or less physical in nature. The point is that *all* known material processes and explanatory principles apply to organisms, while only a limited number of them apply to nonliving systems. . . . Biology, then, is the science that stands at the center of all science. It is the science most directly aimed at science's major goal and most definitive of that goal. And it is here, in the field where all the principles of all the sciences are embodied, that science can truly become unified.[252]

In commenting on this passage, Ernst Mayr voiced the hope that "biologists, physicists, and philosophers working together can construct a broad-based, unified science that incorporates both the living and nonliving world. . . . Paradoxical as it may seem, recognizing the autonomy of biology is the first step toward such a unification."[253] To a skeptical biochemist it seems curious that chemists were omitted from this assembly of intellectual talent, and the "paradoxical" claim suggests that the kind of biology for whose auton-

249. Bernard (1879), vol. 1, p. 248.
250. Kluyver and Donker (1926); see also van Niel (1957).
251. Haynes (1989), p. 2.
252. Simpson (1964b), pp. 106–107.
253. Mayr (1988), p. 21.

omy Mayr strives would exclude the many scientists who may, in the future, advance biological knowledge through the study of the chemistry of life.

In questioning Mayr's seemingly anti-biochemical bias, I am conscious of the profit I derived from the study of his magisterial treatise (1982) on the development of biological thought. Together with Theodosius Dobzhansky's *Genetics and the Origin of Species* (1937), Mayr's book provided my introduction to modern evolutionary theory. For example, I learned of such achievements as those of Sergei Chetverikov, who combined theoretical insight and the naturalist's craft to provide empirical evidence, based on both field observation and laboratory experiment, in support of his view that changes in biological populations are largely a consequence of the action of natural selection on a store of genetic variability generated by mutation.[254] I also was led to read Mayr's outstanding research reports on the evolution of bird populations,[255] and his important writings on the problems of biological classification, from which I learned that, as in other fields of scientific endeavor, there are many questions in that field which remain to be examined through future empirical study. It seems to me, however, somewhat premature for a distinguished evolutionary biologist, writing as a philosopher, to advocate the unification of biology, let alone of all the sciences, at a time when evolutionary theory is still a matter of active scientific and philosophical discourse and not infrequent controversy.[256]

In particular, I do not find in Mayr's sharp distinction between "proximate" and "ultimate" causes, or in his treatment of the question of "determinism" versus "indeterminism" in biological evolution, or in his discussion of the matter of "purpose" (whether it be called teleology or teleonomy), much to chart the course of future research on the mechanisms of Darwinian evolution. It is not enough, I believe, to write of "adaptation" or "fitness" or "predation" or any of the other aspects of the evolutionary process without parallel practical efforts to study such biological phenomena not only by field observation but also, when possible, by laboratory

254. Adams (1980).

255. See Mayr and Short (1970), Mayr (1976), and references therein to Mayr's earlier ornithological studies.

256. See Mayr and Provine (1980).

experimentation.[257] Mayr quotes, with approval, the statement by Pantin that "in astronomy, in geology, and in biology observation of natural events at chosen times and places can sometimes provide information as wholly sufficient for a conclusion to be drawn as that which can be obtained by experiment."[258] The validity of this statement cannot be questioned, but it is another matter to say that "observation in biology has probably produced more insights than all experiments combined."[259] It would be shortsighted, in my opinion, for evolutionary biologists to dismiss the recent efforts of chemically-minded biologists and biologically-minded organic chemists to study experimentally such important features of the evolutionary process as courtship and mate selection, defense against predation, or the mechanisms of olfaction.[260] I also recognize that much (perhaps most) of the actual course of biological evolution, like the course of human history, or even the historical development of scientific knowledge, cannot now be "explained" in terms such as those used in what Mayr calls "functional biology." [261]

I believe, therefore, that for the growth of biological knowledge, it is less important to seek "unity" or "autonomy" than to ask what central problems can be approached fruitfully through scientific practice, by the choice of organisms found to be suitable for such investigations, and by the use of any new methods or instruments which are available or can be devised. At present, the central problems which present the greatest challenges appear to be those of the mechanisms of embryonic development, the role of non-chromosomal inheritance, and the processes in the neurophysiological control of bodily functions, including those associated with learning and thought. All these problems are intimately linked to the mechanisms of biological evolution, whether at the level of the natural selection of phenotypes or at the level of the interaction of organisms with each other and with their environment. The past history of the

257. For a discussion of adaptationist explanations, see Resnick (1989). Beatty and Finsen (1989) have discussed the propensity interpretation of fitness.

258. Pantin (1968), p. 17.

259. Mayr (1982), p. 32.

260. See, for example, Prestwich and Blomquist (1987) and Snyder et al. (1988).

261. Scriven (1959); Edsall (1982). For recent writings on the analogy of the historical development of science to biological evolution, see Toulmin (1967) and Hull (1988).

development of biology, and of its interplay with chemistry, suggests that the further elaboration of evolutionary theory lies not in philosophical speculation about "unity" but rather in the utmost diversity of biological and chemical thought and practice. I believe that the continued growth of reliable knowledge about biological organisms is less likely to be furthered by intellectually satisfying discourse about such matters as "reductionism" than by the intellectual stimulus provided by theoretical insight and empirical discovery to engage in innovative and skillful scientific practice. In offering this opinion, I recognize that what is at issue is the thorny philosophical problem of the nature of scientific explanation. As Stephan Körner wrote, this problem is evident

> . . . when we consider disputes between groups of scientists about two alternative theories, neither of which from the point of view of logical propriety and predictability is superior to the other. In such disputes it is often the case that one group regards one of the competing theories as being the 'real explanation', while rejecting the other as 'intellectually unsatisfactory'. What is required for a 'real explanation' depends on the scientist's conception of what constitutes a 'good' theory–e.g. that such a theory must be mechanistic, stochastic, teleological, etc. These conceptions or attitudes are articulated by normative, regulative or programmatic principles for the construction of theories and are not captured by any analysis of the formal structure of theories and their relations to experience. They are rooted not only in logic and observation, but also in the scientist's view of the world as whole, i.e. in his metaphysics.[262]

262. Körner (1976), p. 90.

CHAPTER FOUR

Approaches to the History of the Biochemical Sciences

During this century, the acceleration in the growth of biological and chemical knowledge, and in the use of that knowledge in medical, agricultural, and industrial practice, has been paralleled by the growth of a community of professional historians whose principal interest lies in the study of these developments and their antecedents.[1] In addition, much of the extensive literature about the history of the chemical and biochemical sciences has been provided by men and women who had been practitioners in some sectors of these fields of research. I will consider later in this chapter the relative merits of some of these contributions, but now only call attention to the contrast between the approach of these "insiders" to that of the succession of professional historians whose philosophical, sociological, or cultural views about the nature of scientific thought and practice have influenced their accounts of the history of the sciences. Some present-day historians of science may consider mention of this contrast to be irrelevant to their various current purposes as scholars or as "discipline builders," but I would suggest that their recent controversies about questions relating to the methodology of their own research discipline bear some of the marks of past tensions arising from the role of scientific workers as historians of their specialties.

1. See Kragh (1987); Olby et al. (1990). For the early development of the professional history of science in France, see Paul (1976), and in the United States, see Thackray (1980) and Reingold (1986).

The issue was squarely put, during the middle years of this century, by George Sarton, then the leading advocate in the United States of professionalism in the history of the sciences. In 1938, he took a dim view of the scientist "who has become sufficiently interested in the genesis of his knowledge to wish to investigate it, but has no idea whatever how such investigations should be conducted and is not even aware of his shortcomings . . . He generally lacks the humility of the beginner, and publishes his results with blind and fatuous assurance."[2] This statement, whose partial validity I will not dispute, must be seen, however, in relation to such declarations as "The main duty of the historian of science is the defense of tradition" or "We owe gratitude to the benefactors of the past, in particular the great men of science who opened new paths, and also the lesser men who helped them, for we are standing on their shoulders."[3] My views regarding the task of the historian of science, whether he or she be classified as an amateur intruder or a professional, are rather different from those expounded by Sarton.[4] These differences arise only partly from my previous activity as a professional biochemist, and more importantly from my opinions about the nature, methods, and purposes of historical scholarship.

I cannot do justice to the vast literature on the methodology of historical investigation,[5] and will only acknowledge my special debt to two small books, published during the 1950s. One was by the Dutch historian Pieter Geyl, who taught me that history and myth are inextricably linked, and that the task of the critical historian is to do battle with myths. Also, Geyl warned that a modern historian should be "on his guard against . . . intellectually satisfying schemes which may hedge in or distort the view. . . . He will not too readily identify a period with an idea: behind the idea he will look for the unruly, struggling men. Behind the anonymity of a class, of a nation, of a sect he will search for various shadings, for individual peculiarities."[6] The other book was the English translation of Marc

2. Sarton (1938), p. 469.

3. Sarton (1952), pp. 15–16. For the "shoulders of giants," see Merton (1965).

4. For an evaluation of Sarton's role in the development of the study of the history of the sciences, see Thackray and Merton (1972).

5. A useful recent bibliography may be found in Marwick (1989), pp. 409–419.

6. Geyl (1955), p. 80. See also Fischer (1970). "Myth" was defined by Marwick (1989), p. 401, as "a version of the past which usually has some element of truth in it, but which distorts what actually happened in support of some vested interest."

Bloch's unfinished *Apologie pour l'Histoire, ou Métier d'Historien.* According to Bloch, the purpose of historical scholarship is to deepen understanding of the present through close and critical study of human activity in the past. As he put it, "Misunderstanding of the present is the inevitable consequence of ignorance of the past. But a man may wear himself out just as fruitlessly in seeking to understand the past if he is totally ignorant of the present."[7] Another passage which struck a responsive chord deals with the examination of historical data: "For a long time, critical techniques were practiced, at least with any consistency, almost exclusively by a handful of scholars, exegetes and connoisseurs. Writers engaged in historical works of the high-flown sort scarcely bothered to familiarize themselves with such laboratory exercises, far too detailed for their taste, or even to take their results into account. Now, as Humboldt put it, it is never good for chemists to be afraid 'of getting their hands wet.'"[8]

For a scientist who embarks on a project relating to some aspect of the historical development of his or her specialty, the "laboratory exercises" involve, first of all, the painstaking collection and critical examination of the factual material considered at the start to be relevant to the problem under investigation. If the project happens to be, as in my case, a study of the interplay between chemical and biological thought and practice since about 1800, one encounters in previously published writings a vast variety of accounts of persons and events associated with scientific theories, observations, experiments, and methods. Many of the items of historical data presented as "facts" turn out, during the course of further study, to be fictions, and questions then arise as to reasons for such lapses in historical accuracy. It may be found that some errors were derived from the repeated copying of incorrect information provided by a widely used basic reference source, but more usually the reason is that an earlier writer did not know (or ignored) other data inconsistent with those accepted as being "true" or even invented spurious "facts" (myths) to buttress a preconceived notion. Consequently, in his efforts as a historian, a scientist must search assiduously for accurate information about the careers of all the individual scientists encountered in

7. Bloch (1954), p. 43. In its original version, this passage reads: "L'incompréhension du présent naît fatalement de l'ignorance du passé. Mais il n'est peut-être pas moins vain de s'épuiser à comprendre le passé, si l'on ne sait rien du présent."

8. Ibid., pp. 85–86. For an admirable recent biography of Bloch, see Fink (1989).

his survey, including the lesser folk who achieved little fame.[9] This means not only the examination of what these scientists may have written or were reported to have said about their researches, their personal lives, and their social relationships in published books and articles, or in surviving diaries and private correspondence, but also the consultation of public records held in state, city, company, and church archives. Much of the autobiographical material based on the recollections of both noted and less famous scientists is likely to require careful scrutiny, for, no matter how honest its purpose, memory is uncertain and may reflect self-serving efforts to reassert reputation and to judge previous allies and rivals in less than objective terms. Nor can the historian expect objectivity in biographical writings, many of which have been either acts of homage (often bordering on hagiography) or iconoclastic denigrations of such laudatory accounts. Nevertheless, all these sources are valuable because they provide information which needs to be checked for its scientific and historical accuracy and, more important, because they may reveal how individual scientists chose to portray the role they had played in the development of their area of scientific inquiry. Moreover, although the principal interest of the historian may be in the scientific education and research achievements of a particular individual, material about that person's other interests (sometimes the dominant ones) is likely to put into better perspective his or her personal qualities as a scientist.

For the historian of an area of modern scientific inquiry, however, such scrutiny must, in my view, be based on the informed and critical examination of what scientists chose to write about their scientific work and thought in professional journals, semi-popular books, and published lectures, as well as in laboratory notebooks. In particular, I emphasize the necessity of the informed study of such original writings, and the requirement that the reader must have acquired an adequate knowledge of the common national language in which they were written and also of the changes in the meaning of scientific terms and concepts since they were used in these writings. For this purpose, it is useful to examine the widely used textbooks and laboratory manuals published throughout the historical period under study. Obviously, the items of scientific

9. For example, see Fruton (1982c). A supplement to this bio-bibliography was also published in 1985 by the American Philosophical Society.

information so assembled will not be equally interesting in relation to a historical problem which is larger than the question of the accuracy of the documentary record. The nature of the problem thus provides the basis for the selection of those data which are deemed by the historian to be significant, and the interpretation of their place in the historical process under study becomes the main task of the critical historian. Such interpretation requires, in my opinion, an appreciation of the characteristic methodology of the scientists under study, and especially of the changes in that methodology (even in the work of an individual scientist) in response to new empirical data and new hypotheses as well as to the invention of new methods and instruments or significant improvement of older ones. In my view, the methodology of a particular area of research in the chemical and the biological sciences develops during the course of that research, and the ever-changing diversity of approach should be more important to the historian than his or her commitment to some idealization of the "scientific method."

I know that these opinions regarding historical scholarship reflect the practice of many of the leading professional historians of the modern chemical and biological sciences, but some of those more inclined to sociological approaches to the understanding of the growth of scientific knowledge have disparaged the value of such "internalist" history, and have given primary attention to the social, political, and institutional factors in that development.[10] Although the strife, especially intense during the 1970s, within the community of historians of science over the issue of the "internalist" versus the "externalist" attitudes appears to have abated, the issue still remains, because scientific research is indeed a social process. Some professional sociologists of science, notably Robert Merton, have contributed much to the understanding of that process through meticulous scholarship and historical insight.[11] In my opinion, however, such understanding is not likely to be promoted by disregard of what scientists wrote about their scientific work and thought, and by considering solely their roles as entrepreneurs and politicians. My own studies on the social and institutional relationships among members of chemical and biochemical research groups, and with their patrons, have convinced me that such relationships are inti-

10. Reingold (1981).
11. Merton (1957, 1973).

mately linked with the course of scientific investigation. In my opinion, therefore, social historians who, either through ignorance or disdain, choose to disregard the detailed record of theory and practice within an area of scientific inquiry are likely to produce myths no less egregious than those produced by some "internalist" scientist-historians.

As I indicated in the previous two chapters, another source of mythology about the interplay of the modern chemical and biological sciences has been the philosophical approach which seeks to interpret the history of these sciences principally in terms of a succession of theoretical concepts, comparable to those in the development of mathematical physics from Newton to Heisenberg. There can be no doubt that metaphysical ideas have played a large role in the elaboration of theories of chemical structure or of the organization of biological function. Thus, the emergence of the modern biochemical sciences since 1800 has been marked by a competition between two styles of molecular explanation of biological phenomena, one in which molecules were regarded as units of physical motion while in the other they were viewed as units of chemical reaction.[12] Concepts such as those of the chemical atom and molecule, or of chemical valence and structure, were questioned on philosophical grounds by some leading scientists until organic chemists who followed Kekulé showed that they could apply these metaphysical theories to the artificial synthesis of materials identical in every measurable respect with natural constituents of biological systems.[13] In biology, the ancient philosophical debate over mechanism versus vitalism was transformed during the middle of the nineteenth century by the experimental achievements of physiologists such as Hermann Helmholtz, Adolf Fick, and Claude Bernard, and, as I indicated in the preceding chapter, the quality of that debate has been further altered during this century by the growth of empirical knowledge in biochemistry and in genetics. A feature of the recent discourse which is especially relevant is the seeming indifference to the past "internal" history of these fields, evident in the writings not only of some philosophers of biology but also of some of the distinguished scientists who, during the period 1945–1960, provided much of this new empirical knowledge.

12. Fruton (1976).
13. Rocke (1984).

On Historians of Chemistry

In recent decades it has become fashionable within the community of professional historians of science to disparage the efforts of chemists as historians, on the ground that their accounts are tainted with "internalism" and "Whiggery." An example of such criticism is the review by Arnold Thackray of the fourth volume of James Riddick Partington's *A History of Chemistry*.[14] According to Thackray, although this volume is "firmly imbedded in the old, and by now weary tradition of chemical history by chemists for chemists," it "immediately makes clear the enormous debt everyone interested in the history of chemistry owes to Partington. One can only marvel at the labour, the erudition, the patience that has gone into this work." After calling attention to the failure of such chemist-historians to "appreciate or reflect the new climate of historical scholarship,"[15] which Thackray attributed to the influence of Paul Tannery, George Sarton, Alexandre Koyré, Karl Popper, and Thomas Kuhn, as well as to selected British general historians (Herbert Butterfield, Robin George Collingwood, Edward Hallett Carr), Thackray deplored Partington's "obsessive concern with facts" and found "little sign of the trained historian's discrimination."[16] This negative opinion was echoed a few years later in the statement that "Partington's refusal to induce any general conclusions from his copious data has left us with an encyclopedia which cannot be read as narrative . . . written by chemists for chemists, and . . . limited in historical insight."[17]

Such judgments invite several comments. First, I reiterate my belief that the practice of sound historical scholarship requires the collection, organization, and evaluation of the factual data relevant to the historical problem under investigation, as best as the validity of these data can be ascertained. My admittedly selective reading of books by British general historians, including those by Butterfield, Collingwood, and Carr, suggests that Thackray's appeal to their authority does not reflect the opinion of other leading lights of "the new climate of historical scholarship" in Britain, let alone in other nations, on the subject of historical "facts." For example, the polit-

14. Partington (1961–1970).
15. Thackray (1966), p. 126.
16. Ibid., p. 129. See also Williams (1966), p. 203.
17. Fisher (1973a), p. 53.

ical historian Geoffrey Elton, in his comments about Carr's views, wrote:

> Mere dates or names—lists of kings or popes—are supposed to be more lowly than tables of trade statistics or the arguments to defend a political philosophy. This is a game not worth the playing, for the peasants of this hierarchical society are as vital to it as its princes. Without the simple details of chronology, genealogy and historical geography, history would have no existence. . . . [It] is not a question of interpreting fact but of establishing it, and the differences resulting are likely to be in the degree and depth of knowledge, no more.[18]

Whatever may be said about the deficiences of Partington's "encyclopedia," and they are many, it has been of inestimable value not only to chemists but also to present-day historians of modern chemistry, as have other books of this kind, such as that of Ida Freund on chemical composition or those by Carl Schorlemmer, Edvard Hjelt, Carl Graebe, and Paul Walden on organic chemistry.[19] To be sure, such books are not free of bias, in particular of nationalism and opportunism. For example, Adolphe Wurtz, writing on the eve of the Franco-Prussian War, stated that "chemistry is a French science. It was established by Lavoisier, of immortal memory," and Paul Walden, writing in the time of Hitler, wrote that there was "a specific German chemistry, a specific style of German chemical research."[20] A critical and chemically-educated historian, whose aim is to understand and interpret the development of chemical thought and practice, will recognize such prejudice and take it into account in using the books of these writers as sources of historical data. Likewise, such a historian will recognize the bias of compendia of statistical data such as that recently prepared by Thackray and his colleagues. In the preface to this book, it is stated that "this volume does not offer history. Instead it provides certain elements —indicators—that may be useful to individuals interested in the history of American chemistry and chemical industry, and sugges-

18. Elton (1967), p. 60. See also Marwick (1989), pp. 193–198, for a critical evaluation of Carr's writings about historical "facts."

19. Schorlemmer (1894); Freund (1904); Hjelt (1916); Graebe (1920); Walden (1941). See also Weyer (1973, 1974) and C. A. Russell (1988).

20. Wurtz (1869), p. I; Walden (1935), p. 2.

tive for policy."[21] It may be questioned, however, to what extent such statistical data will be "suggestive for policy"; in my opinion, their value for the understanding of the contributions of American chemists to the growth of knowledge is likely to be far less significant than that of the data provided by "internalist" chroniclers and historians who collected and interpreted "facts" about the scientific education and achievements of individual American scientists and their research groups.

The repeated criticism that books such as that of Partington were written by scientists for other members of their particular specialty is frequently accompanied by the assertion that professional historians of science should address their writings not to scientists but to a "wider audience," presumably meaning historians, sociologists, philosophers, and members of other non-scientific professions. I find in this counsel a tinge of opportunistic pandering to potential patrons, as well as a tendency which evokes recollection of Macaulay's *Essays*, which, in Geyl's words, were "too much attuned to the expectations and prejudices of the contemporary public, expectations and prejudices that we no longer share; some of these, indeed, we could never have shared, because they were so exclusively and intolerantly English."[22] Whatever the merit of reliable accounts addressed to non-scientific audiences, I also believe that a major purpose of professional historians of the various modern sciences ought to be, through meticulous scholarship, the enlightenment of the practitioners in these sciences. If members of the present generation of historians continues to reject this aim, and are indifferent to the technical details in the development of a major area of modern scientific research, they will, in my opinion, do a disservice both to the advancement of their profession and to the education of future scientists who perhaps may later be inclined to become patrons of their enterprise.

Although the principal focus of this chapter is on the historiography of the modern chemical and biological sciences, these remarks are also intended to apply to some of the recent historical writings on other areas of scientific inquiry, ranging from quantum theory to twentieth-century scientific contributions to medical practice. For example, in his valuable "internalist" article on the development of

21. Thackray et al. (1985), p. xiii.
22. Geyl (1958), p. 30.

modern quantum theory, Michael Redhead noted that "the more recent history has largely been dealt with by practising physicists, rather than by historians," and that

> an important controversy in the historiography of quantum theory has concerned the relevance of wider cultural influences, where some authors have sought to identify the source of new scientific ideas. These writers have probably overstated their case. The quantum theorists were driven by great intellectual ingenuity in confronting what most of them, arguably correctly, regarded as objective problem situations arising from the interplay of theory and experiment in atomic physics.[23]

Numerous other examples can be cited of valuable recent historical scholarship which has illuminated the development of particular areas of scientific inquiry through critical studies of their "internal" history. I call attention especially to the writings of Rachel Laudan on the successive transformations of the conceptual framework and methodology of geology.[24]

The field of the history of modern medicine as a scholarly pursuit was established in this century by physician-historians such as Karl Sudhoff, Charles Singer, Henry Sigerist, Erwin Ackerknecht, and Owsei Temkin. Their books were written for the enlightenment of physicians and of others who valued such qualities as clinical or experimental skill, and who wished to learn more about the interplay of medical practice and scientific research, as well as about the development of the social role of medicine. However, as Leonard Wilson wrote in 1980,

> A striking feature of some recent publications in the history of medicine is the degree to which they restrict themselves to medical or public health aspects of social history. They tend to neglect questions of clinical medicine, of the biology of disease, and of science, even when such questions had a direct bearing on the particular historical subject with which they are concerned. The result is incomplete and sometimes severely distorted history. . . . The social history of attitudes toward disease

23. Redhead (1990), p. 458; see also Forman (1971) and Hendry (1980).
24. Laudan (1983, 1989) and Newcomb (1990).

> and doctors is a valid and fascinating subject of inquiry, but it does not comprehend the whole of the history of medicine. In a strict sense the social history may not even be medical history. If such social history be considered medical history, it is medical history without basic medical sciences and clinical methods and concepts; that is, it is medical history without medicine.[25]

One of the most frequent, and often justifiable, accusations leveled against chemist-historians has been that of "Whiggery" or "Whiggism." Many of the myths in the history of chemistry have indeed arisen from this tendency to select from the tortuous development of chemical theory and practice only those events which can be seen to have contributed to the attainment of a later stage of that development. However, Rupert Hall, a distinguished historian of science and an admirer of Butterfield, who in 1931 introduced the catch phrase "Whig interpretation of history," has noted that Butterfield himself, in his later writings, succumbed to his own purported fallacy.[26] As argued by Hall (and others), the growth of scientific knowledge since 1700 is consistent with some idea of a progress whose historical course can best be discerned from a later vantage point. In the study of the historical development of modern chemistry, and its relation to biology during the nineteenth and twentieth centuries, a detailed knowledge of the present state of these areas of scientific inquiry is, in my view, indispensable for the understanding of the past, and the counsel to historians of science that the less they know about the present, the better their perception of the past, is therefore misguided.[27]

In his 1966 critique of Partington's volume, Thackray also deplored the failure among chemist-historians of recent times to take full account of the writings of philosophically-minded historians of science from Tannery to Kuhn. I doubt whether this homage to the founders and innovators in the professional study of the history of the sciences would be as unqualified today as it may have been during the 1960s. There can be no doubt that Tannery, Koyré, and Sarton did much to promote the emergence of a cadre of professional

25. Wilson (1980), pp. 6–7. See also Rothschuh (1980) and Warner (1985).
26. Hall (1983); Bédairida (1987), p. 346. See also Mayr (1990).
27. See, for example, Russell (1984) and Harrison (1987).

historians of science, especially through their insistence on the necessity for rigor and precision in historical scholarship. Their philosophical bias, however, inclined them to exalt scientific theory over observation, experiment, and craftsmanship. Thus, in his criticism of Alistair Crombie's important studies on medieval science, Koyré wrote:

> The great scientific revolutions of the twentieth century—as well as those of the seventeenth or nineteenth—although naturally founded on the discovery of new facts—or on the impossibility of verifying them—are fundamentally *theoretical* revolutions whose result was not to unite more effectively the "data of experience," but to gain a new conception of the deep reality which underlies these data.[28]

Indeed, Thackray might have been well advised to include among the proponents of the "new history of science" scholars such as Crombie who, during the early 1960s, defined its problems in terms which reflected the recognition of the intimate linking of scientific theory and practice with each other and with the social factors in their development.[29] As for the other worthies cited by Thackray, I do not believe that a historian of modern chemistry should, in Sarton's words, "defend tradition" but rather that he or she should examine critically all the "traditions" as they replaced each other during the course of the interplay of theory and practice. In my opinion, the inclusion of Popper was inappropriate because of his assertion that the historical process is unknowable as a real object, and because I find his "logic of scientific discovery" to be ahistorical and therefore irrelevant to the scholarly investigation of the history of chemistry. My admiration of Kuhn's historical scholarship and insight is unbounded, and is not lessened by my agreement with Körner's view that Kuhn's description of a conceptual transition, such as that from Newton's to Einstein's universe, "is incomplete in that it fails to account for the division of scientists and historians of science into two groups: one, to which Kuhn belongs, which

28. Koyré (1956), p. 42. For Koyré, see Gillispie (1973).
29. See Crombie (1963).

regards the transition as revolutionary; the other, to which Einstein belongs, which regards it as normal progress."[30]

This comment is of particular importance in relation to historical writings about the development of modern chemical theory and practice. Thus, the approach to the question whether a particular transition in chemical thought was a "revolution" or "normal progress" or "reform" may be influenced by the extent of the historian's knowledge of the scientific "facts" as recounted not only by other historians but also by chemist-historians such as Partington. Thus, what was proclaimed as a "revolution" in chemistry (or in any other area of scientific inquiry) by some of the participants in a particular development may be seen later in different perspectives by both chemist-historians and professional historians of chemistry.[31] There can be no doubt that some chemist-historians have introduced considerable mythology into the historical record through such devices as tables of scientific genealogy in which Justus von Liebig is the primogenitor.[32] Also, the tradition of identifying an instrument or a method in terms such as Liebig condenser, Bunsen burner, or Claisen rearrangement has given generations of chemistry students a rather warped view of the role of these men in the development of chemical knowledge. To these failings one may add the obsession with "origins" and the uncritical acceptance of statements which assign discoveries to particular individuals or the uncritical attributions of prescience to noted scientists.[33] The exposure and correction of myths arising from these practices require the historian to know not only the relevant details of past "internal" history, but also what happened afterward, perhaps up to the present.[34]

It seems to me that in their justifiable skepticism about the data and interpretations presented in the writings of chemist-historians, some professional historians of chemistry have gone too far in their dismissal of the past efforts of scientists to describe the development

30. Körner (1986), p. 13. To Körner's comment should be added the later statement by Kuhn (1989), p. 49, that "my past work has often invoked discontinuity, and my present paper points the way to a significant reformulation."

31. See, for example, Cohen (1985), Gillispie (1985), Rocke (1985), and Perrin (1987).

32. Fruton (1990), p. 11. See also Zuckerman (1987).

33. See Goldwhite (1975) and James (1985) for examples. See also Witkop's introduction to the reprint of Emil Fischer's *Aus meinem Leben* (1987), pp. xxi–xxii, for "Fischer's prediction of 'genetic engineering.'"

34. Merton (1961); Sandler (1979); Hull (1979).

of their specialty. For example, in 1966 Pearce Williams concluded a review of several books as follows:

> The historian of science need not be a scientist, but he must know the language of science. Science, to him, is a tool, like French and German, without which he cannot understand his material, but he must look at this material from the perspective that can only be gained by a rigorous training in history.[35]

Apart from the question of which philosophical and methodological approaches constitute "rigorous training in history," the statement that to the historian "science is a tool, like French or German," seems somewhat at variance with Williams's earlier assertion that "the purely scientific aspects of the nineteenth century are overwhelming in their number and complexity. The person who ventures into this period without a sound scientific training takes a very real risk of either missing a point entirely or seriously misinterpreting the evidence."[36] After calling attention to the "presentist" tendency of scientist-historians, Williams stated:

> The life of science in the nineteenth century was an intensely vital and controversial one, involving strong personalities whose non-scientific attributes often influenced the course of science. . . . Put another way, the essence of the history of science is biographical and one wants to know the total person to whom a new theory is due if the genesis of his ideas is to be understood. These ideas do not always arise from objective nature but rather from the the idiosyncratic viewpoint of unique individuals.[37]

Obviously, the historian of science who ventures into the twentieth century faces even greater challenges, and it may be that chemist-historians can help to lessen the risks mentioned by Williams. Also, in his emphasis on the biographical approach to the "genesis of ideas," Williams reflected a view which subordinates scientific practice to theory and craftsmanship to intellect. As I noted in

35. Williams (1966), p. 204.
36. Ibid., p. 200.
37. Ibid.

Chapter 2, such a view limits understanding of the historical development of the modern chemical sciences; indeed, Williams's outstanding biography of Michael Faraday provides exemplary evidence of the fruitful interplay of practical skill and ingenuity with theoretical insight in the career of a great chemist.[38] Moreover, it may be questioned whether the "essence" of the history of the modern sciences can be fully captured solely through the study of the life, work, and thought of notable individuals. Although I agree that critical biography is the most important tool in historical research, and that professional historians of the sciences have contributed much to the correction of myths about such individuals, I believe that closer attention should also be given to the complexity of the relationship between the development, during the nineteenth and twentieth centuries, of particular areas of scientific inquiry and the emergence of separate institutional disciplines. Both scientist-historians and professional historians, especially those of sociological inclination, have, in my opinion, introduced considerable mythology into the history of the transformation of scientific specialties.

On Historians of the Biochemical Sciences

Because I was once a research biochemist, a teacher of biochemistry, a coauthor of a textbook of biochemistry, and a participant in the effort to loosen the dependence of biochemistry on its role in the education of American medical students, I have been aware of the distinction which the critical historian must make between the growth of knowledge about what has been called the "chemistry of life" and the social and institutional factors which have affected that development.[39] After the emergence of biochemistry, during the 1930s, as a central discipline linking the chemical and biological sciences, and the subsequent emergence, during the 1960s, of molecular biology as its historical successor, I undertook to describe, in some detail, the interactions of various kinds of scientific workers, ranging from pharmacists to physicists, in the study of several scientific problems which were considered during the nineteenth century to be important, and which had been more clearly defined by

38. Williams (1965). See also Cohen (1956, 1990) on Benjamin Franklin.
39. Fruton (1951); Fruton and Simmonds (1958).

the middle of the twentieth century.[40] These problems concerned the nature of enzymes and proteins, the material basis of heredity, biological oxidation, the pathways of metabolism, and the chemical organization of biological systems. The book also included an introductory chapter dealing with the social and institutional factors in the development of the interplay of biology and chemistry, and I also stated that the book was not intended to be a comprehensive history of biochemistry.[41] One such book, on the history of physiological chemistry, by the Austrian biochemist Fritz Lieben, had been published in 1935; it still represents a valuable source of historical data, as perceived from the perspective of the medically-oriented biochemists of the early 1930s.[42] Another began to appear in 1972, as a series of volumes by the Belgian biochemist Marcel Florkin.[43]

In 1975 Robert Kohler, a professional historian of biochemistry who had written valuable "internalist" papers on the work of Eduard Buchner and of Rudolf Schoenheimer, published a survey of the historiography of the field. In that survey, Kohler discussed the books by Lieben, Florkin, and myself, and stated that "the scope of all of these works is limited almost exclusively to the subject matter that would interest biochemists, namely the history of discoveries. All transmit more or less uncritically the accepted mythology of the trade and its progressivist historiography."[44] After this echo of Thackray's criticism of Partington, Kohler devoted most of his attention to my book, "whose intended lesson to historians is clear: what is important for history is the discovery of correct facts and theories, and the historian is to chronicle these. Larger questions of social organization, the ideals of the discipline, and historical interpretation come second."[45] This caricature of my "intended lesson" was embedded in a mélange of compliments ("many valuable insights," "impressive scholarship") and criticisms. Among the latter was the view that my book represented "vertical" history, and that "by historians' standards of the history of ideas, Fruton's history

40. Fruton (1972).

41. Ibid., pp. 1–21.

42. Lieben (1935). There had also appeared notable accounts of the historical development of knowledge about cell respiration by the biochemist David Keilin (1966) and about muscular contraction by the biochemist Dorothy Needham (1971).

43. Florkin (1972, 1979). For a critical review, see Teich (1980).

44. Kohler (1975), p. 277. (Reprinted by permission of Kluwer Academic Publishers.)

45. Ibid., p. 278.

of biochemical ideas falls short of the mark. It is 'internal,' that is parochial."[46] These remarks were followed by a series of statements about "what is wanted" in a proper history of biochemistry. According to Kohler, "There are at least three established historical modes that could be used in writing a history of the sort envisioned above: social or sociological history, intellectual history, and cultural or general history."[47]

It should already be clear that I do not believe that dogmatic assertions about "established historical modes" offer useful guidance to sound historical scholarship, except for the requirement that the historian should strive for the utmost accuracy in the presentation of the available data that he or she considers relevant to the problem under study. I do not question the value of any of Kohler's supposedly "established modes" of historical inquiry, but what is missing from his precepts is an explicit acknowledgment of that requirement, and the recognition of the pitfalls of allowing preconceived ideas to determine the selection and interpretation of historical data. Therefore, in evaluating the merit of Kohler's survey of the historiography of biochemistry, it seems appropriate to consider his own book in this field, published in 1982.[48] Before offering my own opinions, I cite those of three respected professional historians of science. Leonard Wilson wrote:

> Professor Kohler seeks to write the history of biochemistry as a discipline, by which he means the history of biochemists as a professional group. From his standpoint biochemical knowledge and ideas were not influential in the emergence of biochemistry as a discipline except insofar as biochemists used their knowledge as a political tool to advance their professional interests. . . . [He] cites the view of the sociologists Joseph Ben David and Avraham Zloczower that scientists exploit intellectual opportunities not to obtain new scientific knowledge, but to obtain "specialized chairs, institutes, and stable budget lines" (p. 4). Kohler accepts the opinion of the political scientist Yaron Ezrahi that scientists form an interest group whose members used scientific knowledge as a political tool to gain public sup-

46. Ibid., p. 270.
47. Ibid., pp. 279–280.
48. Kohler (1982).

> port for science. Such writers, Kohler notes "share a conception of scientists as social actors in specific institutional contexts. They all use the language of competition, entrepreneurship, and resource management to understand the changing political map of scientific disciplines" (p. 6). Evidently such writers do not feel a need to use changes in scientific knowledge to understand changes in scientific disciplines, and Kohler makes their conception of "a political economy of science" central to his analysis of the emergence of biochemistry as a discipline in the United States.[49]

Wilson then called attention to the omission from Kohler's account of relevant historical data relating to Philip Shaffer, Edward Kendall, and John Jacob Abel. In regard to Abel, Wilson wrote: "Kohler's obliviousness to the historical impact of the isolation of insulin is representative of his refusal to consider the influence of growth and change of scientific knowledge on many scientists whom he discusses, or mentions in passing."[50]

A more sympathetic treatment of Kohler's book may be found in the review by John Servos, who wrote that Kohler

> . . . sometimes seems more interested in defending his claims than in testing them; indeed portions of *From Medical Chemistry to Biochemistry* read much like a lawyer's brief. Perhaps it is possible to understand the growth of a discipline entirely in terms of social, political, and economic circumstances, but to be convincing, such an interpretation must be compared with others and judged according to criteria such as scope, depth, and simplicity of explanation. Kohler does not consistently do this, and as a result elements of his argument, such as his efforts to derive the character of particular research programs from the social context in which they were hatched, seem strained or incomplete.[51]

49. Wilson (1983), p. 462. For a critique of Zloczower's approach, see Turner et al. (1984).

50. Wilson (1983), p. 464.

51. Servos (1983), p. 274. See also Servos (1990) for an admirably balanced account of the emergence of physical chemistry in the United States.

To these two opinions should added that of Frederic Holmes, who found that "Kohler's account, based as it is on extensive archival sources, is convincing. There is reason to be wary, however, about the reliability of some of the factual detail."[52] Holmes then cited some inaccuracies in Kohler's account of the career of Hans Krebs, and concluded his review as follows:

> By reconstructing the institutional contexts within which biochemistry developed, Kohler has made progress towards an eventual description of the contours of that subject. A modern scientific discipline is, however, an enormously complex cluster of individual leaders and followers, schools, subspecialties, literature, methods, concepts, traditions, careers, solved and unsolved problems, evolutionary and revolutionary events, as well as a set of institutional organizations. To understand the interplay of all these factors is a formidable task, the magnitude of which social historians of science have scarcely begun to appreciate.[53]

In my own review of Kohler's book, I welcomed his effort to contribute to the program of historical study he envisioned in 1975, and called attention to items of faulty scholarship not mentioned by Wilson, Servos, and Holmes.[54] Apart from these deficiencies and Kohler's penchant for ill-founded generalizations, which render his book a less than reliable source of historical information, I deplore what I believe to be his two main assumptions, namely that biochemistry is, and has been, a "biomedical discipline," and that the concept of a socially determined professional discipline (however it may be defined) is more important for the understanding of the growth of scientific knowledge than the more complex concept of scientific research as an individual and collective effort to elucidate particular problems within a particular area of scientific inquiry. I do not question the importance of the study of the social and institutional factors in the historical development of the modern natural sciences, but to an accused "internalist" who later ventured into

52. Holmes (1982), p. 779.
53. Ibid., p. 780.
54. Fruton (1983); Fruton (1990), p. 113.

some aspects of the sociology of the chemical and biochemical sciences, Kohler's book, in his words, "falls short of the mark." I disagree, therefore, with the opinion of Jane Maienschein that

> Kohler provides an unusual example of what a good and original institutional history can be. His work is complex and occasionally dry with detail, but it nonetheless yields an important perspective on a specialty within American biology and chemistry. Kohler staunchly maintains that scientific disciplines are political institutions that rise and fall as other political institutions do. They do not reflect essential, fixed categories in nature, and thus they may vary greatly from one setting to another. For example, he argues that American biochemistry's close alliance with medicine made theoretical changes in molecular biology more difficult. For Kohler, institutional factors strongly limit and direct science, a view very different from that of the intellectual historian. But he does not claim that ideas are irrelevant.[55]

For a student of the history of biochemistry, this favorable appraisal raises questions regarding the standards of historical scholarship now being applied within the profession of the history of biology. Apparently, for Maienschein, a historical work can be "good" even if it contains a multitude of errors of commission and omission, and involves the selection, from the vast caldron of available historical data, of only those which fit a preconceived notion.

On Scientific Disciplines

In considering the merit of Kohler's approach, as compared with that of other "established modes" of historical scholarship or of the "internalist" approaches of scientist-historians, it is necessary, in my opinion, to recall that the word "discipline" has several meanings. Indeed, the definition of the term, or such related terms as "interdisciplinarity," has been the subject of much recent discussion among historians, sociologists, and philosophers of the sciences.[56] I

55. Maienschein (1985), p. 151.

56. Bechtel (1986), pp. 3–52, 77–105. I cite this reference only because it contains a useful bibliography and because it focuses on the history of the biochemical sciences.

do not propose to enter that arena, nor to examine the value of alternative terms such as "paradigm" or "research networks." Rather, I prefer to view the question of scientific disciplines from the perspective of one who, through experience as a scientist and as a historian of my specialty, sees the modern scientific enterprise as a complex human activity in which one may discern at least four prominent features, which I will denote by means of the generally familiar words "research," "teaching," "administration," and "publicity." In my opinion, when a historian writes about the development of a particular scientific discipline, he ought to specify which of these (or other) kinds of human activity were considered to constitute the characteristic features of the development of that discipline. In particular, I believe it to be necessary to give close attention to the objectives of individual scientists who have engaged in one or more of these activities and the relative importance such scientists attached to these kinds of activities at successive stages in their professional careers, as well as the extent to which an assessment can be made of the influence of these various activities on the subsequent changes in the conceptual, practical, institutional, and public status of particular areas of scientific inquiry designated as disciplines.

It should also be noted that the word "discipline," in the sense of a human activity, is related to the words "vocation" and "profession." As applied to the scientific life, the first of these two terms has usually carried the idealistic connotation of a "calling" or commitment to the advancement of knowledge within an area of scientific inquiry, whereas the second has denoted, more realistically, the social category of the activity in which scientists earn their living.[57] Clearly, both terms may be applied in varying degree to the majority of the people who have been labeled "scientists," but the changes in the balance between the ideal of vocation and the practicality of profession may be reflected in changes in the social character of a discipline.[58] Some recent writers have expressed a measure of nostalgia, not unmixed with mythology, in deploring the loss of the values they have attributed to a lost "Golden Age" of an area of

57. See Tondl (1979) and Pelikan (1984).

58. By "social character" I mean particularly the extent of the adherence to such "norms" as have been put forward by Merton (1973), pp. 256–278, 286–324. See also Ziman (1984), pp. 81–90.

inquiry as a consequence of its professionalization and its increasing dependence on financial support from public agencies.

For the historian of the growth of knowledge in the chemical and biological sciences, clearly the most important activity has been, and continues to be, personal research, perhaps with the help of one or two students, apprentices, or technicians but, more recently, by the leadership of a larger, closely directed group of junior associates. Frequently, such research has involved the close collaboration of pairs of investigators of approximately equal academic status, as in the cases of Liebig and Wöhler, Bunsen and Kirchhoff, or Crick and Watson. The main objectives of such research have been, and continue to be, the investigation of particular problems of interest to the research scientist, the discovery of new empirical knowledge, and the invention of new or improved methods and instruments, as well as the formulation and testing of new hypotheses relating to these problems. The ideas, empirical data, and techniques developed in various settings (laboratories, museums, field expeditions) by the individual persons or groups have been, and continue to be, communicated in the form of publications (or "pre-publications") to others interested in the same or similar research problems. Within the overarching general disciplines now termed chemistry and biology, whose boundaries have been defined by the gap between the knowledge at a given time between the "living" and the "non-living," there have emerged a succession of sub-disciplines which, in time, either were elevated to the status of disciplines (for example, genetics) or were absorbed into the framework of either established or emergent disciplines, as in the case of mineralogy.[59]

Thus, in terms of the activity of scientists in research, the definition of a particular discipline involves statements about the nature of the problems of common interest to the participants in that activity. Such definitions are likely to be different from those based on other features of the scientific life, for example in the case of a member of a university chemistry department who studies the sex attractants of insects. A historian of the modern chemical and biological sciences must therefore take account of differences between "research disciplines" and "institutional disciplines." Unless the historian chooses to disregard the research activity of the members

59. Laudan (1987).

of an institutional discipline, he or she will be obliged to evaluate the significance and quality of the research efforts of particular individuals, and may find that the evaluation by other, perhaps more prominent, contemporary workers in the same field was different from that of later investigators in that area of inquiry or from that of scientist-historians. Moreover, results considered to be important and reliable at the time of their publication may be quickly adopted as part of the groundwork for further research and, if such later research is fruitful, these earlier contributions may lose any privileged status they may have once been accorded. By the same token, some research efforts thought by leading contemporaries to be trivial or to be lacking in adherence to the standards of their time may, in the light of later discovery, be judged to have been pioneering contributions to the emergence of a particular discipline. I believe, therefore, that historians of the chemical and biological sciences need to take more account of the *quality* of these contributions, as judged by the standards at the time they were made. The historian's definition of a discipline, in the sense of a collective inquiry into a set of related scientific problems, thus requires an appreciation of the changes in the boundaries and accepted standards within an area of research before judgments can be made about whether a particular contribution was significant or trivial in the development of knowledge within that discipline. To evade this challenge, and to seek explanations of the historical transformation of an area of research solely in terms of the activities of scientists as teachers, administrators, or publicists is likely to provide a rather limited understanding of the process whereby such an area became an officially designated discipline.

In some respects, whether or not an individual identified as a scientist was productive as an investigator, all such people have, in one way or another, taught others. This educational activity has taken various forms, among them the authorship of textbooks (often based on university lecture courses) or of semi-popular books or the supervision of practical work (often leading to the publication of laboratory or field manuals). It is not, in my opinion, a valid criticism of such activities to consider them solely as a means of propagating a prevalent dogma. Rather, they should be viewed by the informed and critical historian as important features of the transformation of scientific disciplines. Among the many examples which may be cited, Charles Lyell's *Principles of Geology* (1830–1833) "shaped geologists' conceptions of their discipline."[60] In my own area of historical inquiry, the successive textbooks of Fourcroy and Berzelius, Dumas and Liebig, Lehmann, Kühne, Hoppe-Seyler, and Hammarsten; down to those of Fürth, Bayliss, Bodansky, Gortner, and Baldwin, Fruton and Simmonds, and Lehninger; and, most recently, those of Stryer and of Singer and Berg, offer to the historian an opportunity to examine the ways in which the concepts and methods of the biochemical sciences have been taught during the course of the interplay of biology and chemistry which led to the emergence of the disciplines now included among these sciences.[61]

Such textbooks have served not only as systematic introductions to the state of an academic discipline, as perceived at a given time by individual teachers, but also (in many cases) as sources of information about current research problems. However, a teacher, whether an author or a user of these textbooks in lecture courses, and thus identified with an academic discipline, may no longer have been an active participant in the collective research effort to study one or more of these problems. The principal function of most members of chemical and biological departments of colleges and universities, apart from their own research and the education of future scientists, has been the instruction of the myriads of students who have attended their lectures and performed elementary laboratory work in order to meet the requirements for entry into the professional programs of schools of medicine, public health, pharmacy, or chemical technology. The question thus arises whether historians of the chemical and biological sciences should make a distinction between "research disciplines" and "academic disciplines." I believe that such a distinction might help to clarify the roles played by noted chemists and biologists in the emergence of scientific specialties as separate disciplines. To be sure, many of the great lecturers were, at the same time, leading research scientists. Thus, among the noted nineteenth-century university professors of chemistry, Justus Liebig and Robert Wilhelm Bunsen are still remembered as having been

60. Laudan (1983), p. 99.

61. Fourcroy (1801–1802); Berzelius (1833–1836); Dumas (1837), Liebig (1842); Lehmann (1853); Kühne (1866); Hoppe-Seyler (1877–1878); Hammarsten (1895); Fürth (1912–1913); Bayliss (1924); Gortner (1929); Bodansky (1938); Baldwin (1952); Fruton and Simmonds (1958); Lehninger (1975); Stryer (1988); Singer and Berg (1990).

outstanding teachers, and in this century Peter Debye and Robert Woodward were accomplished lecturers to students. It may well be that, in some of the present-day rapidly developing research specialties, such exemplars are less highly regarded, but I believe that historians of the chemical and biological sciences ought to give more attention to influential teachers, whether or not such scientists were also great investigators or successful entrepreneurs.[62]

In many instances, a scientist who achieved distinction in a particular research specialty had received little formal instruction in that field, but acquired such education through later association with experienced workers in that specialty. An example which readily comes to mind is that of the physicist Max Delbrück, who learned how to work with bacteriophages from the biologist Emory Ellis at the California Institute of Technology. Also, from the late nineteenth century onward, many physicians who later engaged in biochemical research had gone to famous university chemical laboratories, such as that of Emil Fischer. Moreover, such postdoctoral education was often gained from teachers who were not affiliated with a university faculty but were located at research establishments such as the Rockefeller Institute in New York and Princeton, the National Institutes of Health in Bethesda, the Pasteur Institute in Paris, the Lister Institute or the National Institute for Medical Research in London, the Carlsberg Laboratory in Copenhagen, or one of the Kaiser-Wilhelm (later, Max-Planck) Institutes in Germany. To these famous non-academic institutions should be added the agricultural experiment stations, hospitals, museums of natural history, and research laboratories of large chemical and pharmaceutical companies, where young scientists were given an opportunity to begin research in association with more experienced workers.

The third type of activity which has been important in shaping the development of particular disciplines within the chemical and biological sciences is the one I have called "administration." This activity includes the relationship, within an institutional unit of research and instruction, of the administrative head of that unit to his or her junior colleagues. I have discussed elsewhere the various styles of the leadership of relatively large research groups in the chemical and biochemical sciences, and have suggested that youth-

62. See Geison (1978) on Michael Foster; Holmes (1989a) on Justus Liebig. See also Olesko (1988).

ful talent was more likely to flower in university departments in which the administrator acted as a counselor rather than a quasi-military dictator.[63] The differences in the style of leadership involved many personal and institutional factors. If, as in the case of some twentieth-century biochemists, the administrative head aspired to win a Nobel Prize, he was likely to collect a departmental faculty largely composed of members of his own research group. Other leading biochemists, especially in American medical school departments during the years 1910–1930, chose to staff their departments with people whose research and teaching activities were likely to gain the approval of the dominant clinical professors. One of the consequences of the latter practice was an emphasis on problems of human nutrition and clinical chemistry, rather than on the more "highbrow" research problems studied at that time by European biochemists.[64] Both nutrition and clinical chemistry are today important biochemical disciplines, but they stand apart from those which have developed since the 1930s. Also, during the period 1910–1930, many American medical school departments of physiological chemistry were headed by men whose principal professional ambition appears to have been preferment within the administrative structure of their institution. Such departments were generally small, and staffed by conscientious teachers fully occupied in the formal instruction of first-year medical students.

It should be noted, however, that some of the heads of biochemical departments in Germany, Britain, and the United States which, under their leadership, became outstanding centers of research and instruction appear to have been rather modest in their personal ambitions and not very energetic in promoting the institutional status of their departments. Among these people I would include Franz Hofmeister at Strassburg, Frederick Gowland Hopkins at Cambridge, and Hans Thacher Clarke at Columbia University. It would seem, therefore, that "discipline building" has been not only a matter of institutional politics but also one of vision, tolerance, and, perhaps above all, the ability to recognize and encourage scientific talent. Moreover, although people such as Hopkins and Clarke may not have been effective entrepreneurs, they exerted considerable influence through the respect they had won from enlightened patrons

63. Fruton (1990).
64. Kohler (1982), p. 252.

such as Walter Morley Fletcher and Edward Mellanby of the British Medical Research Council and Warren Weaver of the Rockefeller Foundation.[65] I must leave to future historians of the biochemical sciences the question of the extent to which the more recent, and more generous, patronage of these sciences has been influenced by the counsel of scientists possessing the personal qualities of Hopkins and Clarke. I hope that future historians of science will examine the changes, during the latter half of this century, in the administrative and advisory roles of leaders in the biochemical sciences vis-à-vis the managerial bureaucracy in universities and research institutes, and especially in their relations to the dispensers of public or private financial support. In particular, one may ask to what extent the social character of the biochemical sciences in the United States has been affected by the proliferation of research specialties in these sciences, and by such factors as the use (or misuse) of the "peer review" system of major federal granting agencies.[66] In raising these questions, I readily acknowledge the fact that the development of knowledge in the biochemical sciences, since at least 1800, has been intimately linked to the efforts of prominent advocates of a research specialty to promote and, when challenged, to defend the group interests of that specialty. In my opinion, what is at issue is whether those social historians of science who have projected into the nineteenth and the early twentieth century the "political economy" of the present-day biochemical sciences may have been unduly influenced by the current social climate of these sciences.

These remarks about the administrative activities of some university biochemists in the promotion of their discipline are intended to suggest that, depending on their personal aspirations and on their institutional setting, these people appear to have held widely different views about the place of the biochemical sciences not only in their own institution, but also in the growth of scientific knowledge. Such differences in attitude present to the historian of science the problem of assessing the relative importance, in the development of a discipline, of the role of a scientist as an administrator, as compared with his or her role as an active researcher and educator. In addition, it may be asked whether, and in what ways, these attitudes have changed with the increased public prestige of the biochemical

65. Kohler (1976, 1978).
66. Cole et al. (1981).

specialties in recent decades as a consequence of widely recognized successes in research, and as reflected in the vast increase in financial support from governmental agencies, private foundations, and industrial concerns. Moreover, since the development of a biochemical discipline has been, and now is more than ever, an international enterprise, British and American historians need to take into account the differences and changes in the organization and control of universities and research institutes, not only in Britain and the United States but also in other nations such as Germany, France, Russia, Japan, Switzerland, or Sweden.[67]

Individual scientists and research groups in the modern chemical and biological sciences have gained a reputation within their field of inquiry through research articles in specialized professional journals and in some widely read periodicals such as the British *Nature* or the American *Science*. Such communication of new theoretical and empirical knowledge is a distinctive feature of scientific activity and is essential in the formation of separate disciplines. This topic will be discussed in the next chapter. Equally important, however, have been the responses of scientists in distantly related disciplines or of historians, philosophers, and educated public officials to the claims of protagonists of a "new" specialty that it deserves a distinctive status among the sciences. In the case of the chemical and biological sciences, such claims have usually emphasized the fruitful application to medical, agricultural, or industrial practice of the knowledge provided by those identified as practitioners in that specialty. Other spokesmen have stressed the contribution of their specialty to the "unification" of the sciences. The advocacy of such claims is not a recent phenomenon, for famous nineteenth-century chemists, notably Justus Liebig and Wilhelm Ostwald, were energetic publicists.[68] In this century, with the accelerated growth of the scientific population and the multiplication of specialties within the chemical and biological sciences, many noted scientists have appeared in the public arena as champions of their particular discipline. As before, it has been a matter of using whatever media were available to communicate their message to a wider public audience.

67. Many books and articles have appeared in recent years on these topics. I have learned much from those by Chubin (1976), Hünemörder and Scheele (1977), Burchardt (1978, 1980), Elias et al. (1982), Jarausch (1983), Johnson (1985a, 1985b), Clark (1987), and especially Becher (1989).

68. Fischer (1955); Turner (1982).

Because of the public acceptance of Nobel Prizes in physics, chemistry, and physiology or medicine as marks of the highest distinction in the natural sciences, the writings of recipients of these awards have been accorded special attention.

For many scientists the word "publicity" has had, often justifiably, a pejorative connotation, but it may also be argued that the widest possible dissemination of knowledge about new developments in the sciences is a social good. To be sure, in the hands of journalists, primarily beholden to their employers, public statements by scientists have often been distorted so as to increase their "news value." Thus, in 1906, the world press announced that, owing to the work of Emil Fischer on polypeptides, the artificial synthesis of natural proteins was just around the corner.[69] In more recent times, the great achievements of Arthur Kornberg, Robert Sinsheimer, and Mehran Goulian in their studies on the biosynthesis of DNA offer another example of the public response to an important advance in the biochemical sciences.[70] If public memory of the promises offered in such publicity is fleeting, the repetition of these promises tends to result in increasing institutional and public recognition of the disciplines with which these scientists are associated.

On the "Origins of Molecular Biology"

In this section I consider some of the recent writings about the emergence, after the Second World War, of molecular biology as a prominent research discipline. Many of the achievements associated with this field were mentioned in the preceding chapter, and I shall not dwell on them here, except to note the ways in which they have been described and interpreted both by noted participants (or their acolytes) in the research during the period 1945–1970, and by professional historians of science.

One of the recurrent themes in the recent discourse on the origins of molecular biology has been the description offered during the 1960s by the famous X-ray crystallographer John Kendrew, the first editor of the *Journal of Molecular Biology*, which began to appear in

69. See Kautsch (1906). For Emil Fischer's reaction to the newspaper reports, especially the one attributed to Emil Abderhalden, see Fischer's letter to Carl Oppenheimer, 27 January 1906 (in the Fischer collection at the Bancroft Library, University of California, Berkeley).

70. See Greenberg and Singer (1967).

1959. He suggested that molecular biology represented the meeting ground of two research programs, which he denoted as "informational" and "conformational" (or "structural").[71] At that time, the principal areas of research embodying these two programs were bacterial and phage genetics and X-ray crystallographic studies on proteins and nucleic acids. As the noted phage geneticist Salvador Luria later put it:

> Molecular biology deals with questions of molecular structure, and therefore is biochemistry; but it is not the classical biochemistry that emerged earlier in the twentieth century out of the concerns of medical, agricultural, and industrial researchers. Molecular biology is genetics because it deals with genes, their functions, and their products; but, in contrast with classical genetics, it has dealt mainly with organisms such as bacteria and viruses rather than peas, maize or fruit flies, whose study had established the classical rules of genetics.[72]

In an earlier statement, Luria defined molecular biology as "the program of interpreting the specific structures and functions of organisms in terms of molecular structure."[73] The historical questions raised by such statements by distinguished leaders of both the "structural" and "informational" schools of molecular biology have been discussed extensively by other prominent scientists, by well-informed journalists, and by professional historians, sociologists, and philosophers of the sciences. I have derived from their books and articles much valuable information, but my treatment of some of the questions may differ in emphasis and interpretation.

First, I wish to dispose of what I consider to be the relatively minor question of who introduced the term "molecular biology." There is no doubt that the term was used in 1938 by Warren Weaver in an annual report of the Rockefeller Foundation.[74] Nor is there doubt that the X-ray crystallographer William Astbury used the term during the 1940s and early 1950s.[75] Moreover, during the first years of this century, the noted Belgian plant physiologist Léo Errera, who

71. Kendrew (1968).
72. Luria (1984), pp. 83–84.
73. Luria (1970), p. 1289.
74. Weaver (1970a); see also Kohler (1976).
75. Astbury (1952); see also Hess (1970).

had studied with Felix Hoppe-Seyler, published a book entitled *Physiologie Moléculaire*. In my view, however, questions about the "originators" of terms later adopted as the names of scientific disciplines are of doubtful value in aiding historical understanding, and the answers have often introduced an element of mythology into the history of the sciences. For example, it has been stated that the term "Biochemie" was introduced by Carl Neuberg in 1906, although it had been used frequently during the nineteenth century, as in the title of a book by the Austrian chemist Vincenz Kletzinsky, published in 1858.[76] In this relatively undistinguished book, Kletzinsky separated biology into three disciplines: biochemistry, biophysics, and biomorphology. The emergence during the first half of the twentieth century of biochemistry and of biophysics as prominent scientific disciplines has, in my view, little to do with whatever prescience later historians may have wished to attribute to this neglected chemist. By the same token, I consider it to be an idle historical exercise to argue whether it was Weaver or Astbury, or someone else, who introduced the term "molecular biology" that was subsequently adopted as the designation of a prominent research discipline. Rather, I find more worthy of attention the fact that the statement of a now-forgotten nineteenth-century chemist about the nature of the biological problems of his time appears to have elicited no discernible changes in the status of biochemistry or biophysics as official institutional disciplines within the biological sciences. By the same token, though in a reverse sense, it seems to me a matter of greater historical interest to examine more closely the roles of scientists as researchers, educators, administrators, and publicists, as well of those of professional historians or journalists, in the description of the programs of research advocated by Weaver and Astbury and later considered to be distinctive features of the new discipline of molecular biology. In a thoughtful, but brief, recent account of the emergence of molecular biology as a research discipline, the historian Robert Olby wrote:

> Rather than searching for the immediate antecedents of modern molecular biology, let us explore the broad conception of a 'molecular' biology from which has emerged and evolved the mo-

76. Kletzinsky (1858); for an obituary notice for Kletzinsky, see Hoswell (1882). See also Fruton (1976), p. 332.

> lecular biology we know today. The manner in which this modern molecular biology has been rendered more complex and in some of its aspects transformed, is a chapter which awaits the historian.[77]

It is in the spirit of this appeal that I offer some comments about several of the problems likely to be encountered in the historical enterprise outlined by Olby.

Although future historians of molecular biology will gain much factual information from the various books and many journal articles about the initial stages in the development of this discipline, they will, I believe, be obliged to subject these accounts to more searching criticism than that already accorded them.[78] I hope they will not be constrained, in Geyl's phrase, by "intellectually satisfying schemes which may hedge in or distort the view," for they will find in many of these accounts a prevalence of myth, occasionally mixed with vainglory. Apart from a tendency to construct a pantheon of the "founders," the effort to emphasize the revolutionary character of the transition from a "classical" biochemistry or genetics to a new unified discipline has, in my view, obscured rather than illuminated a complex historical process. As I noted in the previous chapter, that process has long included the influence of thought and practice in research disciplines such as organic chemistry, physical chemistry, and physics on the development of biological research disciplines such as cytology, physiology, and microbiology, as well as medical specialties such as pharmacology or immunology. I therefore find it to be rather arbitrary to draw a sharp distinction between the concept of "molecular structure" that developed before the public appearance of molecular biology and the concept of "molecular structure" that has emerged from the outstanding achievements of X-ray crystallographers in their studies on proteins and nucleic acids. Moreover, what Luria termed "classical biochemistry"

77. Olby (1990), p. 519.

78. The books written by non-scientists which attracted the most attention were those by Olby (1974) and by Judson (1979); for critical reviews, see Cohen (1975), Fruton (1975), Teich (1975), and Abir-Am (1985). There has been a succession of books by some of the scientists considered to have been the founders of molecular biology, or collections of essays about these scientists. The most famous of these books is, of course, the one by Watson (1968); others were written or edited by Cairns et al. (1966), Lwoff and Ullmann (1979), Luria (1984), and Crick (1988).

did not emerge solely "out of the concerns of medical, agricultural, and industrial researchers." On the contrary, many of the leaders (among them Jacques Loeb, Emil Fischer, Frederick Gowland Hopkins, Otto Warburg, Albert Jan Kluyver, Otto Meyerhof, and Hans Krebs) in the emergence, during the period 1900–1950, of a "general biochemistry" voiced aspirations to "explain" biological phenomena in the language of chemistry and physics and, through their own contributions and their influence on several generations of people who called themselves biochemists or biophysicists, created a climate of scientific opinion favorable to the search for "molecular" descriptions of the structure and function of living things. It is, I believe, erroneous to claim that because DNA, RNA, and proteins are chemical substances of crucial importance in the propagation and survival of living things, the study of their three-dimensional structure has a status different from that of the investigation of the three-dimensional structure of other biochemical constituents such as the brain hormones or vitamin B_{12}. Future historians may perhaps ask whether such notable figures in the emergence of molecular biology as Frederick Sanger and Gobind Khorana were not, after all, outstanding chemists, and why the word "molecular" was preferred to the word "chemical" in the description of their achievements. Was it because prominent spokesmen for the new molecular biology had sought to leap from physics to biology without setting foot in chemistry?

As for the distinction Luria drew between "classical" genetics, based on studies on "peas, maize, and fruit flies," and more recent work on "bacteria and viruses," I doubt whether his statement can withstand critical historical examination. Much has already been written about the work of Oswald Avery's group which led to the identification of the pneumococcal "transforming factor" as a DNA, and why Delbrück's "Phage Group" preferred to hail the later experiments of Alfred Hershey and Martha Chase as decisive evidence for the DNA nature of the gene.[79] More could be written, however, about the studies during the years 1920–1940 on bacterial metabolism and their role in the emergence of molecular biology.[80] Indeed, Jacques

79. See Pollock (1970); Stent (1972); Wyatt (1974, 1975); Hotchkiss (1979); Luria (1984), pp. 86–87; McCarty (1985); Russell (1988b).

80. See Rahn (1932), pp. 270–294; Dubos (1945); Stephenson (1949). Among recent historical writings on bacterial biochemistry, those of Kohler (1985a, 1985b) are especially valuable.

Monod's research on the induction and repression of enzyme formation in bacteria began with his famous doctoral dissertation on bacterial growth.

A prominent feature of writings about the development of bacterial and viral genetics has been the emphasis on the transfer of "information" in biological processes.[81] The future historian of the origins of molecular biology will, I hope, examine critically the purported role of information (or communication) theory, as popularized by the mathematical prodigy Norbert Wiener, who during the Second World War had studied such problems as the control of anti-aircraft fire. Afterward, he sought to apply his mathematical approach to communication and to control (which he termed "cybernetics") to biological problems.[82] In writing about Wiener's book on this subject, one of his biographers stated:

> It has contributed to popularizing a way of thinking in communication theory terms, such as feedback, information, control, input, output, stability, homeostasis, prediction, and filtering. On the other hand, it has also contributed to spreading mistaken ideas of what mathematics really means. *Cybernetics* suggests that it means embellishing a non-mathematical text with terms and formulas from highbrow mathematics. This is a style that is too often imitated by those who have no idea of the meaning of the mathematical words they use. Almost all so-called applications of information theory are of this kind.[83]

Although the mathematics of information theory, as further developed by Claude Shannon and John von Neumann and also popularized by Leo Szilard and Louis Brillouin, appears to have had little application in biological research, the language they introduced was eagerly adopted by those engaged in the study of the genetics and metabolism of bacteria and viruses.[84] In addition to many of the words mentioned above, such terms as code, message, and noise acquired a biological meaning. Although biologists now use such

81. See, for example, Sajet (1978) for a French version, based largely on the writings of Lwoff (1962), Monod (1970), and Jacob (1970).

82. Wiener (1948). See also Shannon and Weaver (1949).

83. Freudenthal (1976), p. 347.

84. See, for example, Umbarger (1956), Yates and Pardee (1956), and Atkinson (1965).

terms largely as metaphors, the future historian may be well advised to examine more closely the ways in which the "intellectually satisfying schemes" offered during the 1950s and 1960s, and based on the ideas of information theory, influenced empirical research at that time. There is no doubt that, before the advent of molecular biology, there were speculations about a possible "copying" mechanism in the replication of genes.[85] Nor is there any doubt that the idealized conception of a "genetic code" for protein synthesis, as proposed by the astrophysicist George Gamow and developed by Francis Crick,[86] stimulated important empirical discovery. As I noted in Chapter 2, valuable theories of this kind also led to the recognition of new complexity. For example, during the 1950s it became fashionable among biochemists to speak of the alignment of "activated" amino acids on a nucleic acid "template" and the formation of a polypeptide chain in a subsequent non-enzymatic "zipper" process.[87] This effort to introduce simplicity was soon replaced by the recognition that, as in the study of then-known biochemical processes, a complex and coordinated assembly of specific enzymatic catalysts is involved. The question may also be raised whether the historical problem of the use (and misuse) of information theory in the development of molecular biology after 1945 is comparable to the problem of the use (and misuse) of thermodynamic theory in biological thought during the first half of this century.

Much has been written about the significance of the entry, during the 1940s and 1950s, of mathematical physicists such as Max Delbrück into biology, but less consideration appears to have been given to the distinctive contributions of many experimental physicists and of inventors of new physical instruments (for example, the electron microscope, the scintillation counter, or the density gradient centrifuge) which enlarged the scope of inquiry into biological problems. Thus, there is now an extensive literature, especially since the 1960s, about the influence on the emergence of molecular biology of the publication of Erwin Schrödinger's book *What is Life?*, first published in 1944.[88] It is amply evident that the later pub-

85. See Pollock (1976).
86. Gamow (1954); Crick (1958).
87. See, for example, Lipmann (1954). See also Fruton (1963), pp. 244–249.
88. Jacob (1970), pp. 270–274; Olby (1974), pp. 240–247; Yoxen (1979); Abir-Am (1985), pp. 101–105; Symonds (1986); Perutz (1987); Fischer, E. P. (1987); Moore (1989), pp. 394–404.

licity accorded to this book by some members of the "informational" school of molecular biology contained much mythology, compounded by ignorance and disdain in their attitudes toward studies on the chemical structure of cellular constituents and the functional role of enzymes in the dynamics of biological processes. My personal recollections are perhaps no more worthy of credence than those of the eminent molecular biologists who have been quoted as having said that they were influenced by Schrödinger's book to move from physics to biology, or those who have testified that the book had no influence on their scientific research, but I well remember my reaction when I read it in 1945. As a college student, I had studied with great admiration (albeit with less than complete understanding) expositions of the wave mechanics of de Broglie and Schrödinger. Fifteen years later, after I had become a biochemist, that admiration turned to dismay at the temerity of a noted scientist who could write about "the physical aspect of the living cell" without taking account of the relevant chemical knowledge of his time. It is interesting to look at some of the early reviews of Schrödinger's book. For example, the biochemist and geneticist J. B. S. Haldane wrote:

> Much of the book is devoted to mutation, and the author not merely accepts Delbrück's account of this process, but also writes "If the Delbrück picture fails, we would have to give up further attempts". . . . Actually, I believe that the Delbrück picture will have to be modified profoundly. . . . Perhaps there are more things in chromosomes than are dreamt of even in wave mechanics.[89]

The mythology surrounding Schrödinger's venture into biological speculation is closely linked to that relating to Delbrück's contributions to the emergence of molecular biology. As a skeptical member of the scientific tribe whose efforts Delbrück had publicly derided, and as a student of the history of biological and chemical thought and practice, I searched in the various panegyrics which had appeared between 1966 and 1988 for accounts of Delbrück's distinctive personal accomplishments as a research scientist in the study of

89. Haldane (1945), p. 375; see also Chapter 3, note 13.

bacteriophages.[90] I found much uncritical discussion of his theoretical contribution in 1935 to Schrödinger's biological thought, and accounts of his important collaboration with Emory Ellis and later with Salvador Luria and Thomas Anderson.[91] However, there was little acknowledgment of the fact that Hermann Muller's discovery of the effect of X-rays on the rate of mutation stimulated research during the 1930s by many investigators on the effects of the irradiation of bacteria and bacteriophages. Among them was Eugène Wollman, with whom Luria worked before coming to the United States in 1939.[92] Nor did I find mention of Delbrück's ill-founded excursion into biochemical speculation in 1941 about an autocatalytic mechanism for protein synthesis as a model for the replication of chromosomes.[93] The greatest emphasis was placed on Delbrück's initiative in organizing in 1945 a summer phage course at the Cold Spring Harbor Laboratory for Quantitative Biology. In the American "Phage Group" which emerged from this effort, Delbrück was considered by some sociologists to be both the "intellectual leader" and the "organizational leader."[94] In a volume published in honor of André Lwoff, Jacques Monod wrote as follows:

> In 1946, André and I had the opportunity to attend the memorable Cold Spring Harbor symposium at which the new discipline was to acquire body and soul. The members of the phage church, already established, were all there, with their their three bishops: Max Delbrück, Al Hershey and Salvador Luria. Max was of course the pope. It was he, *Primus inter pares*, who defined the dogma. What Max accepted was faith, what he rejected was heresy . . . [including] all the results obtained by schools before the foundation of the Church or outside it. For example, Max dismissed out of hand the researches of Burnet, Gratia, E. Wollman, den Dooren de Jong on lysogenic bacteria. He refused to accept the reality of the phenomenon and attributed it to unknown impurities.[95]

90. Cairns et al. (1966); Stent (1982); Kay (1985); Fischer (1985); Fischer and Lipson (1988); Symonds (1988).

91. See, for example, Kay (1985), pp. 237–239.

92. Luria (1984), pp. 69–70. See also Lea et al. (1936).

93. Delbrück (1941).

94. Griffith and Mullins (1972); Mullins (1972).

95. Monod (1971), p. 7. My translation does not do justice to the bite of this

This arrogant rejection spurred Lwoff to study the phenomenon more closely, and thus led indirectly to his important discovery of "prophage."

By now, Gunther Stent's oft-cited contribution in 1968 to the historiography of the informational school, with his division of its development into three phases—romantic, dogmatic, academic—has been subjected to sufficiently extensive criticism. The question still remains, however, as to the validity of Stent's reiterated assertions that "not only did the one-dimensional, or informational, school have nothing in common with biochemistry, but its early practitioners were positively hostile to biochemistry" and that "it was through Delbrück that [Niels] Bohr's epistemology became the intellectual infrastructure of molecular biology. It provided for molecular biologists the conceptual guidance for navigating between the Scylla of crude biochemical reductionism, inspired by 19th century physics, and the Charybdis of obscurantist vitalist holism, inspired by 19th century romanticism."[96]

Another question that ought, in my opinion, to interest historians of the origins of molecular biology is presented by the final sentence in Delbrück's famous 1949 paper, in which he expressed the hope that physicists "will create a new intellectual approach to biology which would lend meaning to the ill-used term biophysics."[97] As a research discipline, biophysics, with roots in seventeenth-century iatromechanics, had emerged after about 1850 as a prominent area of inquiry by physiologists concerned with such problems as nerve conduction and muscular contraction. Many members of the British school of physiology, for example Archibald Vivian Hill, considered themselves to be biophysicists.[98] Moreover, other biologists, notably cytologists and embryologists, who sought explanations of the "physical basis of life" found it intellectually satisfying to base their

---

passage. No doubt some social historians of science will find in it an element of Monod's own entrepreneurship, but I hope that they will also compare the scope and quality of the scientific achievements of Monod with those of Delbrück.

96. Stent (1968), p. 790; (1989), p. 13; see also Delbrück (1986), p. 2. See also Judson (1979), p. 62, for Luria's statement about his and Delbrück's attitude to biochemistry (and biochemists). For comments about these statements, see Cohen (1975, 1984). Some of Delbrück's papers of the 1940s also appeared in the *Archives of Biochemistry*.

97. Delbrück (1949), p. 191.

98. Hill (1956).

hypotheses on the "molecular physics" of their time.[99] In such physicalist speculations, the "colloidal" molecules in biological systems were units of mass, motion, and energy, and special attention was given to the properties of these molecules in providing large surfaces for adsorption. The kind of biophysics which emerged from these efforts was rather different from the one espoused by the physiologists who used physical methods to study processes such as vision, muscular contraction, or electrical conduction in the nervous system. Still another area of biophysics emerged from the "medical physics" which developed after Röntgen's discovery of X-rays, with the appearance of diagnostic and therapeutic radiology as clinical specialties. Some of the physicists who had worked during the Second World War on nuclear weapons, and undertook research in radiation biology, identified themselves as biophysicists, but later proclaimed their adherence to the new discipline of molecular biology. For the student of the historical development of both biophysics and molecular biology it is important, I believe, to examine more carefully the interplay, especially during the 1930s and 1940s, of biophysics with biochemistry, physical chemistry, and organic chemistry, especially in the use in these research disciplines of new physical instruments and methods. That development, with growing knowledge of the chemical nature of proteins, enzymes, and nucleic acids, has led to the emergence of molecular biophysics as a distinctive specialty in which molecules are recognized to be units of chemical reaction.[100] I believe, therefore, that historians should consider the possible connection between Stent's reference to "crude biochemical reductionism, inspired by 19th century physics" and Delbrück's usage of the term "biophysics" rather than "molecular biology."

These comments lead me to consider some historical aspects of the other purported component of modern molecular biology, namely "structural" or "conformational" biology, and the rise to importance of the use of X-ray diffraction analysis in the study of the three-dimensional structure of proteins and nucleic acids. There has been, I believe, considerable "Whiggery" in the tendency not only of scientist-historians, but also of some professional historians of sci-

99. Fruton (1976).

100. For a perceptive commentary on the present state of biophysics, see Weber (1990).

ence, to emphasize those features of the development of theory and practice in the study of molecular structure which these writers considered to be significant steps on "the path to the double helix."[101] In my view, these steps should be seen within the framework of the emergence of X-ray crystallography as a distinctive chemical specialty.[102] Particular attention has rightly been given to the role of William Henry Bragg and his son William Lawrence Bragg in founding the British school,[103] but some of the other pioneers in establishing X-ray crystallography as a tool in the determination of molecular structure also merit closer attention.

Among these men was the mineralogist Arthur Hutchinson, who taught at Cambridge for forty years (1891–1931).[104] He was was appointed lecturer in crystallography in 1923, and in 1926 he became professor of mineralogy. As a student, Hutchinson had received his Ph.D. in organic chemistry for work with Emil Fischer in Würzburg, where he had also worked with Röntgen. Upon his appointment to the Cambridge professorship, Hutchinson proposed the establishment of a new lectureship in structural crystallography, and John Desmond Bernal was chosen; William Astbury was one of the other applicants. In 1931, upon Hutchinson's retirement, the administrative control of Bernal's unit was transferred to the Cavendish laboratory of physics.[105] In 1938 Lawrence Bragg became head of that laboratory, and earlier in that year Bernal had been made professor of physics at Birkbeck College in London; the only member of Bernal's group who remained in Cambridge was Max Perutz, who had begun work there in 1936.[106]

Another scientist who merits the closer attention of students of the institutional development of X-ray crystallography is the physical chemist Roscoe Gilkey Dickinson, who was associated with the California Institute of Technology in Pasadena from 1918 until his death in 1945. Dickinson had been encouraged by Arthur Amos

101. Olby (1974).

102. For the early development of X-ray crystallography, see Ewald (1962) and Glusker (1981).

103. Law (1973).

104. For Hutchinson, see Mills (1939) and Smith (1939). For Hutchinson's influence on W. L. Bragg, Astbury, and Bernal, see Hodgkin (1980), pp. 25–26.

105. Hodgkin (1980), pp. 29–36. This admirable obituary notice for Bernal is an indispensable source of historical information. See also Hodgkin (1977) and Goldsmith (1980).

106. Crowther (1974), pp. 301–319.

Noyes, and their most famous student, Linus Pauling, later wrote that "[Dickinson's] determination (with one of his students, A[lbert] L. Raymond) of the structure of hexamethylenetetramine was the first structure determination ever made of a molecule of an organic compound. In a decade he developed the leading American school of X-rays and crystal structure."[107] Some aspects of Pauling's many achievements in that field were discussed in the preceding chapter.

Apart from the seeming neglect by historians of molecular biology of the roles of teachers such as Hutchinson and Dickinson, the selective emphasis on the efforts to use X-ray diffraction methods in the study of macromolecules (polysaccharides, proteins, nucleic acids) has tended to obscure the importance of the travail of the many investigators who developed the art and science of chemical crystallography through studies on the structure of small organic molecules. Among those who had worked at the Royal Institution with William Henry Bragg, Kathleen Yardley Lonsdale and James Monteath Robertson were outstanding this regard. In 1928 Lonsdale established the planarity of the benzene ring through her work on hexamethylbenzene; the story of her life and work, already summarized brilliantly by Dorothy Hodgkin,[108] deserves, in my opinion, more consideration. During the 1930s, one of the most important achievements in the field of X-ray crystallography was Robertson's direct determination of the structure of the metal complexes of the tetracyclic phthalocyanines. Other crystallographers, notably Johannes Martin Bijvoet,[109] can be added to the list of those who, through studies on small organic molecules, contributed much to the solution of the difficult problems involving the interpretation of the complex X-ray diffraction spectra given by macromolecules of biological interest. Indeed, the continued importance of such studies for the further development of X-ray crystallography is evident in the recent recognition of the contributions of Herbert Hauptmann and Jerome Karle.[110] Together with the post–World War II introduction of automation in the collection of data and of computers in their interpretation, such contributions have made X-ray crystallography (along with mass spectroscopy and nuclear magnetic resonance

107. Pauling (1971).
108. Hodgkin (1976); Robertson and Woodward (1937).
109. Groenwege and Peerdemann (1983).
110. Hauptmann and Blessing (1987). David Sayre has also made important contributions in this field.

spectroscopy) the principal physical tools for the determination of molecular structure.[111]

These considerations lead me to question the importance assigned by some historians of molecular biology, notably Robert Olby, to the role of William Astbury in the development of modern X-ray crystallography.[112] As a young biochemist at the Rockefeller Institute during the 1930s, I had the opportunity to meet Astbury and to hear him speak several times about his interpretation of the X-ray diffraction patterns he had obtained with keratin fibers. I found great pleasure in his company, for his enthusiasm was infectious, but I was puzzled by the fact that the structures he derived from these patterns changed on successive presentations. I later recognized that Astbury's espousal of the term "molecular biology" was related to his eagerness to contribute, through his X-ray studies, to the solution of biological problems, and that this aspiration was reflected in his seeming indifference to the necessity of establishing, as accurately as possible, the three-dimensional structure of the monomeric units of which proteins are composed. It is perhaps not unjust to say that this ambition made Astbury a less reliable chemical crystallographer than Bernal.[113] Also, through Bernal's influence on his students, especially Dorothy Crowfoot Hodgkin, Isidor Fankuchen, and Max Perutz, Bernal had a more lasting impact on the later development of his field, and his view of the scope and challenges in that field appears to have been sounder than that of Astbury.[114]

To return to the question of the origins of molecular biology, I offer my own opinion that the principal impetus in its emergence to worldwide prominence during the 1950s and 1960s came not only from the outstanding research achievements of the group led by Perutz at Cambridge, but also from the manner in which that group achieved independent status within the university. Much attention has been given by historians to the role of the Rockefeller Foundation and the British Medical Research Council in aiding the remarkable growth of the enterprise which Perutz had inherited from Bernal in 1938, and I will return to that topic shortly. However, what

111. Hamilton (1970).

112. Olby (1974), pp. 41–70.

113. For a perceptive biography of Astbury, see Bernal (1963). See also Waddington (1969), Witkowski (1980b), and Davies (1990).

114. See, for example, Bernal (1930) for a statement of his views on the place of X-ray crystallography among the scientific specialties. See also Edsall (1983).

has only been sketched thus far is the course of the discussion within the university about the relationship of Perutz's group to the established academic disciplines at Cambridge. I hope that future historians will have the opportunity to study this matter more closely, for I believe that their findings are likely to have more general significance for the understanding of the development of the modern biochemical sciences.

Perutz has written briefly about his early efforts as a lone investigator at Cambridge. After Lawrence Bragg had arrived in 1938, Perutz

> waited from day to day, hoping that Bragg would come around to the Crystallographic Laboratory to find out what was going on there. After about six weeks I plucked up courage and called on him in Rutherford's Victorian office in Free School Lane. When I showed him my X-ray pictures of haemoglobin his face lit up. He realised at once the challenge of extending X-ray analysis to the giant molecules of the living cell, and obtained a grant from the Rockefeller Foundation to appoint me as his research assistant. Bragg's effective action saved my scientific career and enabled me to bring my parents to England, so that they escaped the holocaust.[115]

Soon afterward, with the outbreak of war, Perutz was interned as an enemy alien and shipped to Canada.[116] Upon his return to Cambridge, he resumed his work on hemoglobin and was joined by John Kendrew, his first Ph.D. student. Perutz's academic status was insecure, for he was a chemist on a fellowship from Imperial Chemical Industries, and was working in a physics department on a problem of biological interest. The decisive help came from the biochemist David Keilin, head of the Molteno Institute of Parasitology, who persuaded Bragg to ask the biochemist Edward Mellanby, then secretary of the Medical Research Council, for support of the work of Perutz and Kendrew. Thus, in 1947, there was created at Cambridge the "MRC Unit for the Study of the Molecular Structure of Biological Systems," and as Perutz has put it, "it took nine years before I

115. Perutz (1980), p. 327. See also Olby (1985).
116. Perutz (1989b), pp. 101–148.

thought of shortening its name to 'Molecular Biology.'"[117] By that time other scientists, including Francis Crick, Hugh Huxley, James Watson, and Sydney Brenner, had joined the group, and the hut they occupied next to the Cavendish Laboratory became rather crowded. Accordingly, Perutz

> thought of asking the MRC to build us our own laboratory but I regarded this as premature until the structure of at least one protein had been solved. Even when the solution of Kendrew's myoglobin structure seemed assured I was still doubtful because we were too weak on the chemical side. We therefore stretched out feelers to Fred Sanger, who worked first in a basement room and later in a crammed hut at the Biochemistry School in Tennis Court Road, and asked if he would join us. When he agreed I wrote a letter to the MRC . . . [After the MRC approval in 1958], we still needed a building site and the university's approval. This was far more difficult than winning over the MRC. The top university administrators were strongly against our plan and it took more than two years of negotiating before the university reached agreement with the MRC. We wanted a site in the centre of Cambridge so as to maintain our contacts with colleagues in the university, but in the end we accepted Professor Joseph Mitchell's offer of a site in the future Postgraduate Medical School because no central one was to be had. In March 1962 we finally moved into the lab.[118]

To this report I add my own recollections. Of my many stays in Cambridge, the most memorable one was from August 1962 until June 1963, when I was a guest in Alexander Todd's new chemical palace on Lensfield Road, and made frequent visits to the new Laboratory of Molecular Biology at the foot of Hills Road. It soon became evident that the latter had become the principal center for fruitful research in the biochemical sciences at Cambridge. The presence of Perutz, Kendrew, Crick, Sanger, and their outstanding junior associ-

117. Perutz (1980), p. 328. See also Perutz (1986b) for his tribute to Keilin.

118. Perutz (1980), (This quote first appeared in *New Scientist* Magazine, London.) p. 329. In view of the importance Perutz attached to Sanger's decision to leave the Biochemical Laboratory to join the MRC Laboratory, future historians of molecular biology may find it useful to learn more about the role in that decision of the Cambridge professor of biochemistry, the endocrinologist Frank Young.

ates made the MRC Laboratory a very exciting place, and the excitement reached a peak during the fall of 1962 with the announcements of the Nobel awards to Perutz, Kendrew, and Crick. I believe that this public ratification of the eminence of the MRC Laboratory was the most important factor in the general recognition of molecular biology as a distinctive scientific discipline, and that the subsequent influx of many postdoctoral students, especially from the United States, was a reflection of this opinion.[119] Moreover, it is difficult to escape the impression that the appearance in 1966 of the festschrift for Delbrück was influenced by the public esteem gained by the MRC Laboratory, thus setting off the production of a series of books and articles on the "origins" of molecular biology.

There can be no doubt that the program in quantitative experimental biology initiated by the Rockefeller Foundation in 1932, under the direction of Warren Weaver, greatly accelerated the fruitful interplay of biology with chemistry and physics.[120] During the succeeding twenty years, some $90 million was expended in the form of grants to physical chemists, organic chemists, biochemists, biophysicists, cytologists, and geneticists in the United States, Britain, France, Belgium, Germany, Denmark, and Sweden. However, the claims of some social historians of science that the power of Weaver's purse altered the course of that interplay seem to me rather exaggerated, and tinged with "Whiggery."[121] In his autobiography, published in 1970, Weaver took special pride in having given long-term support to the Cold Spring Harbor Biological Laboratory, to Linus Pauling, and to William Astbury because he then considered "the emergence of the subject now regularly called molecular biology to be one of the greatest developments in the history of science."[122] As one who knew Weaver during the 1940s and 1950s, and who was a grateful recipient of his favor, I also know that his objective at that time was to transcend the institutional constraints which separated chemistry and physics from biology, and to seek individual scientists whose work reflected his aspirations, whatever their departmental affiliations may have been. If a promising young scientist was a member of a university department, a grant to that scientist was made only after the receipt of a formal letter of acceptance

119. Steitz (1986).
120. Kohler (1976, 1977, 1991).
121. See Abir-Am (1982b) and Yoxen (1982).
122. Weaver (1970b), pp. 70–72.

from the president of the institution. This procedure could not fail to enhance the status of the recipient of a Rockefeller Foundation grant, especially when the head of the department was unsympathetic to the recipient's research efforts. To my knowledge, however, Weaver never urged university administrators to establish separate departments of molecular biology. Of course, Weaver made decisions which later proved to be erroneous, and his judgment has been questioned (in my opinion unfairly) with respect to the efforts during the 1930s of the "Biotheoretical Gathering" at Cambridge, led by Joseph Needham and Conrad Hal Waddington, to gain the support of the Foundation.[123] Whatever mistakes Weaver made, he played a significant role in the promotion of the growth of knowledge in the biochemical sciences before these sciences began to receive generous financial aid from governmental agencies. In addition, the program directed by Weaver was, from the beginning, international in scope and was especially important in aiding European scientists after the Second World War.

Any account of the rise to prominence of molecular biology must include the role of the British Medical Research Council, led by a succession of distinguished biologists—Walter Fletcher, Edward Mellanby, and Harold Himsworth—who promoted the growth of knowledge in the biochemical and biophysical sciences.[124] Apart from the relationship of the MRC to the National Institute for Medical Research in London, research units (or laboratories) were established in various universities. For example, in 1934 a unit of "bacterial chemistry" was set up for Paul Fildes in London, and in 1945 Hans Krebs was made head of a research unit at Sheffield. In his autobiography, Krebs wrote:

> In the initial negotations . . . the unit was referred to as 'Unit of Research in Biological Chemistry' but when I saw Sir Edward Mellanby on 12 January 1945, he told me that this designation had only been provisional and was, in fact, undesirable because it would give the erroneous impression that the Medical Research Council was willing to set up Departments of Biological Chemistry or Biochemistry in universities. He asked me to put

123. Abir-Am (1987b). For an account of the efforts of the Cambridge group in chemical embryology, see Witkowski (1987).

124. For Fletcher, see Elliott (1933) and Fletcher (1957).

forward an alternative name which would cover the range of my research activities and yet would be somewhat more specific than 'Biological Chemistry' or 'Biochemistry'. I suggested 'Unit for Research in Cell Metabolism' and this was accepted.[125]

Mellanby, himself a biochemist, was less reluctant to approve the establishment of an MRC Biophysics Unit. In 1947, such a unit was set up at King's College London under the direction of John Randall, the professor of physics. Randall had made a significant contribution to the war effort through his invention of the resonance magnetron, which enabled radar transmitters to be used in aircraft. With support from the Rockefeller Foundation and advice from Bernal, Randall assembled an outstanding group, which included Maurice Wilkins, Rosalind Franklin, and Jean Hanson.[126] I consider Randall to have been the most distinguished member of the group of physicists in Britain and the United States who turned to biological problems after the war, and believe that future historians of science will find in the closer study of his career a clearer view of the place of biophysics in the emergence of molecular biology. By the same token, Mellanby's attitude toward biophysics as an important research discipline and his judgment of the qualities of various biophysicists also deserve more attention.[127] From my conversations with Mellanby during the late 1940s I gained the impression that, like Weaver, he was more interested in the scientific promise of individual investigators in the various research disciplines with which they were associated than in the promotion of the institutional status of these disciplines in universities. Because the organization and control of universities in various nations have been widely different, and have undergone change, the manner in which molecular biology gained recognition as an academic discipline different from country to country. In France the chief support of the Rockefeller Foundation program directed by Weaver went to the Pasteur Institute, in Germany to some of the Kaiser-Wilhelm (later, Max-Planck) institutes, in Sweden to the Karolinska Institute, and in Denmark to the Carlsberg Laboratory, rather than to officially established university departments in these countries. Together with the MRC Labora-

125. Krebs (1981), p. 133.

126. See Wilkins (1987). See also Randall (1951); Olby (1974), pp. 326–331; Wilson (1988). For Hanson, see Randall (1975).

127. See Dale (1955).

tory in Cambridge, these research institutes attracted many foreign (especially American) postdoctoral students who later sought to promote the establishment of comparable research units within the universities where they obtained faculty posts.[128] In the United States this effort was greatly aided, after 1960, by funds from the National Institutes of Health and, more recently, from the Howard Hughes Foundation.

If, as I believe to be the case, molecular biology now represents a complex mélange of interdependent but diverse research tracks which stem chiefly from the no less complex sets of tracks in the biochemistry, biophysics, microbiology, and genetics of the past, the task of the future historians of molecular biology is truly formidable. I hope, however, that they will have a sufficiently wide knowledge, such as that provided by scientist-historians, of the pre–World War II development of the interplay of biological thought and practice with the concepts and methods of chemistry and physics, to examine more objectively than may now be possible some questions relating to the place of molecular biology among the sciences. For example, some distinguished biochemists have suggested that the rise of molecular biology has tended to weaken the traditional demand that students who aspire to do research in the kind of "quantitative experimental biology" envisioned by Weaver receive a thorough grounding in organic and physical chemistry.[129] If this estimate is correct, to what extent has this trend affected the standards by which the quality of research contributions of molecular biologists is now judged? Another question which ought to interest future historians is the extent to which the recent preoccupation with nucleic acids and proteins has led to the neglect of earlier active research tracks, such as those dealing with polysaccharides, lipids, enzyme mechanisms, and metabolic pathways. Also, as is evident from the recent writings of social historians of science, the vast increase in the population of molecular biologists competing within ever-narrower areas of inquiry appears to have reduced adherence to the Mertonian norms of behavior in the relationships within scientific communities. What impact, if any, have such changes in the social climate of a set of research disciplines had on the growth of scientific knowledge?[130]

128. See, for example, Cohen (1986).
129. Kornberg (1987); Chargaff (1989).
130. See, for example, Bollum (1981).

On Scientific Biography and Autobiography

The history of science, to paraphrase the words of Thomas Carlyle, "is the essence of innumerable biographies" as embodied in the stories of the lives, work, and thought of the men and women whom we label "scientists." This view is implicit in the purpose of the eighteen-volume *Dictionary of Scientific Biography* (DSB), now the most valuable single source of reliable information about individuals, from antiquity to the recent past, whom the editors considered to have made significant contributions to the development of mathematics and the natural sciences. Many of the articles were prepared by scientists and, in my opinion, are of more variable quality than those written by professional historians of science. The bibliographies for the articles include many references to autobiographical books and articles by the person under consideration, or biographical material (usually in the form of obituary notices) prepared by scientists familiar with the area of the subject's scientific activity. It is not my intent to criticize the contents of the DSB but, in what follows, to ask some questions raised by this series of volumes, especially as these questions apply to the historiography of the chemical and biological sciences.

One may first ask, what is the proper place of biography (and autobiography) in the practice of the history of the sciences? Is it, as Sarton preached, to acknowledge our gratitude to great men and to defend some concept of the tradition they represented? Or is it subsidiary to the search for generalizations about such concepts as the "scientific method," or individual "productivity" and "creativity," or about the social, institutional, and political milieu at a particular stage of the development of an area of scientific research and education? Is the documentation of the private life of a scientist, in contrast to common practice in literary history, a distraction from a critical study of his scientific work and thought? Should evidence about a scientist's personal behavior affect the historian's judgment about his contribution to the development of scientific knowledge? If one happens to believe that Louis Pasteur and Otto Warburg were two of the more unpleasant characters in modern biochemistry, does the magnitude of their scientific achievements overshadow their seeming deficiencies as human beings? Should it matter to the historian of chemistry that Lavoisier was guillotined because he was a tax collector, not because he was a great chemist? I have no ready answers to such questions, and take refuge in the opinion that a historian's definition of his problem is likely to determine the extent to which accounts of these matters are necessary and appropriate in an exposition of studies on that problem. Also, I believe that, in a sense, scientists have created their own biographies through successive writings about their scientific work and thought. They have written not only to enhance their reputation and to influence the work and thought of their contemporaries, but also in the hope that their writings will survive for a time even after the theories and experiments they describe will have been superseded in the general advance of scientific knowledge. For the critical historian of the sciences, therefore, these writings ought to constitute the principal starting point for an interpretation of the role of a particular scientist in the development of an area of scientific inquiry.

However, as Frederic Holmes has shown in his outstanding studies on the nineteenth-century physiologist Claude Bernard, the eighteenth-century chemist Antoine Lavoisier, and the twentieth-century biochemist Hans Krebs, in order to gain an insight into the investigations of these scientists, it is not sufficient to read their published writings.[131] By means of critical examination of the surviving laboratory records and, in the case of Krebs, oral interviews, Holmes has penetrated more deeply into the individual research styles of these three pioneers in the biochemical sciences than has been customary among either scientist-historians or professional historians of science. For those historians of science who seek facile generalizations, Holmes's "fine-structure analysis" of experiments may be unwelcome evidence of the time-consuming effort, scientific knowledge, and critical judgment required in such scientific biography. It may be that, as in the now-large population of workers in the biochemical sciences, the smaller community of historians of these sciences is afflicted by a competitive and entrepreneurial spirit which is unfavorable to the kind of sustained scholarship evident in the writings of Holmes. I hope, however, that the recognition accorded to his contributions both by historians and by historically-minded scientists may encourage others to follow his example.

The subtitle of Holmes's book on Lavoisier is *An Exploration of Scientific Creativity*, and Holmes has also written several articles about the concept of creativity in scientific research.[132] My admira-

131. Holmes (1974, 1985, 1991); for Bernard, see also Grmek (1973).
132. Holmes (1981, 1986, 1989b).

tion of his historical scholarship is not diminished by my skepticism about the value of that concept in the search for understanding of the development of scientific knowledge. The extensive recent literature about "scientific creativity," largely provided by psychologists or by the new group who call themselves "cognitive scientists,"[133] raises many challenging questions for historians of the sciences. One of these questions derives from the circumstance that a thesis has an antithesis: if there are such persons as "creative" scientists whose styles of work and thought merit special attention, there are presumably other scientists, perhaps in the same area of inquiry, who are, by some definition, "uncreative." Is there not here a tinge of Sarton's views about "great men of science" and perhaps even a taint of "Whiggism"? Or are there degrees of creativity which allow one to decide that, within a group of comparable scientists, some of them have been more creative than others? Gruber has, in my opinion rightly, questioned the value of statistical data, because "the individual disappears."[134] I find it difficult, however, to accept his assertion that "if there is to be a scientific understanding of creative work, it cannot be limited to those few things we may find that some creative people have in common. Instead we must search for a general approach to the description and understanding of unique, creative people."[135] Moreover, I find little illumination in Gruber's discussion of a possible "general approach" in terms of such personal attributes as originality, commitment, purpose, resourcefulness, enterprise, capacity for sustained and difficult work, imagination, or insight, to which he might have added such other qualities as self-discipline, craftsmanship, or ingenuity.

One of the qualities most often attributed to successful scientists is their ability to choose important problems which seemed to them to be amenable to solution. It may perhaps be unfair to suggest that this lesson has not been learned by those who search for a "general understanding" of the uniqueness of individual renowned scientists, but it seems to me that the outcome of this difficult enterprise is far more uncertain than appears to be appreciated by present-day

133. The writings on "scientific creativity" include the books by Taylor and Barron (1963), Krebs and Shelley (1974), Aris et al. (1983), and Wallace and Gruber (1989), as well as many of the journal articles cited therein.

134. Wallace and Gruber (1989), p. 5. See also Gruber (1981).

135. Wallace and Gruber (1989), p. 6. Williams (1985), p. 74, has noted that "creative" has become a cant word, as in "creative advertising."

"cognitive scientists." Indeed, in my opinion, Holmes's admirable books offer the best available evidence for the complexity of the problem of scientific creativity. They provide a corrective not only to romantic accounts such as those of Arthur Koestler, but also to facile generalizations offered by some philosophers and sociologists about the methodology of the chemical and biological sciences. To this I must add that, as a biochemist familiar with the historical background of Krebs's discovery of the ornithine cycle, and with Holmes's writings on this subject, I was not impressed by the report that it was possible to set up a computer program based on Holmes's studies which showed that discovery to have been "produced by a whole sequence of tentative decisions and their consequent findings, not by a single 'flash of insight,' that is, an unmotivated leap."[136] Is it old-fashioned to cling to the view that, for the historian of science, the literary form of critical biography still remains the most valuable means of exploring the manner in which an individual scientist succeeded (or failed) in his or her endeavors?[137] I agree that the biographer should not only study the subject's published writings but also attempt, as Holmes has done, to penetrate into the subject's life in the laboratory. I repeat my hope that others will follow Holmes's example, for the knowledge so gained will further illuminate the interplay of thought and action in modern chemical and biological research. It remains to be seen, however, whether the "fine-structure" study of the research of thirty, or even three hundred, noted chemists and biologists will lead to a "general understanding" of the nature of "scientific creativity." I was introduced to the version of this concept as expounded, for example, by Golovin during the late 1950s in the dismay of the American military establishment at the successful launching of Sputnik.[138] Although that version may be disavowed by present-day "cognitive scientists," it is difficult to escape the impression that the support of their enterprise by the National Science Foundation and by other agencies of the U.S. government has reflected an interest somewhat different from that of critical historians of the sciences. Now that the cold war appears to have abated, will funding by these agencies for studies on "scientific creativity" be justified by the commercial com-

136. Kulkarni and Simon (1988), p. 174. See also Langley et al. (1987) and Qin and Simon (1990).
137. Hankins (1979).
138. Golovin (1963).

petition with Japan? Might it not be more appropriate for historians to replace the much-abused term "creativity" by some less pretentious word, such as "style"? As noted in the previous chapter, Fleck used the term *Denkstil* to denote a collective rather than an individual mode of thought. Although admittedly also susceptible to abuse, the word "style" seems to me to require less dependence on the public recognition, as through a Nobel Prize, of some particular aspects of a scientist's research for the historian's study of the manner in which that scientist worked and thought during the course of his career.

To the elusive concept of scientific creativity should be added that of productivity, as measured in the number of research publications by a scientist. Perhaps in some areas of inquiry, such as modern organic chemistry, it may not be difficult to find some correlation between quantity and quality. In the biochemical sciences, however, such correlation is more uncertain, as is evident from the prodigious output of papers by Emil Abderhalden. In my opinion, the estimable efforts to discern patterns in scientific productivity have raised more questions than they have answered.[139] Nor do I agree with Peter Medawar's statement that, in seeking to learn how scientists did their research,

> it is no use looking to scientific 'papers,' for they not merely conceal but actively misrepresent the reasoning that goes into the work they describe. . . . Nor is it any use listening to accounts of what scientists *say* they do, for their opinions vary widely enough to accommodate almost any methodological hypothesis we may care to devise. Only unstudied evidence will do—and that means listening at a keyhole.[140]

In a previous chapter I questioned some of Medawar's Popperian views about the "scientific method," and I now suggest that this quip, which smacks of donnish talk over port in Oxbridge combination rooms, obscures what I believe to be an important point. When a well-written scientific paper about a particular research problem is read by a person who has done experimental work on the same (or a similar) problem, and has thought about that problem intensively,

139. Zuckerman (1967); Roe (1972); Krebs (1976); Andrews (1979).
140. Medawar (1967), pp. 169–170.

he or she is likely to respond to the most formal report with sympathetic appreciation of the trial and error which attended the author's research. This appreciation will be increased if the reader is also familiar with the earlier research papers by the same author. Perhaps therein lies the most valuable contribution that informed scientist-historians can offer to professional historians of science who seek to gain insight into the development of the styles of research of particular modern chemists or biologists. It may be that "listening at a keyhole" will not only confirm what can be learned from the critical and informed perusal of research papers but will also illuminate further one's understanding of individual styles of scientific work and thought; but for a noted biologist to write that the study of research papers is useless is, in my view, deplorable. I agree with Medawar's statement only to the extent that it refers to what many prominent chemists and biologists have written (or were reported to have said) about their research successes in semi-popular books or articles. There, the complexity of the scientific problem has frequently been oversimplified, the historical background of its investigation has usually been distorted, the contributions of co-workers have been obscured, and self-esteem may have exaggerated scientific talent. Moreover, in reading such accounts it is often forgotten that, in scientific research as in other human activities, innovation is usually preceded by imitation.[141]

Although the achievements of modern chemists and biologists as scientific investigators have usually and, in my opinion, properly taken first place in the recognition accorded them by historians of science, such scientists have also played roles as educators, administrators, or publicists. This diversity of professional activity presents to the historian of the chemical and biological sciences difficult problems in the use of biographical and autobiographical sources. For a dedicated research scientist such as Hans Krebs, these other activities are distractions:

> The majority of scientists are perpetually tempted away from science—to administration, to positions of power, and to teaching. Many are, of course, primarily paid to administer and teach and their employers expect them to give these activities priority. These responsibilities rob scientists of much of the time

141. Sasso (1980).

> they need for research which requires far more time than what is now considered an ordinary working week. Creative artists, as a rule, have all or most of their own time at their disposal. Additional tempting and menacing distractions for the reasonably successful scientist are the invitations to travel, to lecture, to attend conferences and to be feted on these occasions.[142]

Those who knew Krebs will recognize that this attitude was a reflection of the style of his own early professional life, as it was of that of his mentor, Otto Warburg. To the social historian of science who seeks in biography or autobiography evidence of the role of a scientist as an administrator or a publicist, however, successful research is a basis for "legitimizing" the exercise of political power. On the other hand, for the student of the development of scientific education, and the transmission or reform of a tradition within an academic discipline, the role of a scientist as a teacher is likely to assume an importance comparable to that of his or her personal research contributions. Consequently, the biographies of noted scientists are likely to undergo revision, depending on the special interests of the historian. This has been true of some nineteenth-century chemists and biologists: for example, the eminence of the chemist Marcelin Berthelot during his lifetime, and for some years afterward, has faded, as is evident from the valuable recent book by Jacques.[143] No doubt, in due time, myths surrounding the careers of noted twentieth-century chemists and biologists will be subjected to similar critical scrutiny. As before, many of these myths will be found to have their source in autobiographical books and journal articles.

In particular, the future historian of the biochemical sciences will be obliged to consider the merit of the many autobiographical books and articles written by noted scientists during the latter half of this century. The most famous of the books is, of course, James Watson's *The Double Helix*, published in 1968 and republished in 1980 with commentary, reviews, and original papers.[144] I refer the reader to this useful compilation, and will only note that even the most favorable reviews of the original edition raised questions about the histor-

142. Krebs (1976), p. 417. (Quoted by permission of Pergamon Press PLC.)

143. Jacques (1987).

144. Watson (1968, 1980). Among the valuable reviews missing from the latter volume is the one by Bernal (1968).

ical accuracy of Watson's account of the formulation in 1953 of the double-helical model of DNA. As Peter Medawar put it:

> Autobiographies, unlike all other works of literature, are part of their own subject matter. Their lies, if any, are lies *of* their authors but not *about* their authors, who (when discovered in falsehood) merely reveal a truth about themselves, namely that that they are liars. Although it sounds a bit too well remembered, Watson's narrative strikes me as perfectly convincing. This is not to say that the apportionments of credits or demerits are necessarily accurate; that is something which cannot be decided in abstraction, but only after the people mentioned in the book have had their say, if they choose to have it.[145]

The most important person mentioned in Watson's book, apart from himself, is of course Francis Crick, whose subsequent writings about his association with Watson are, in my opinion, likely to be more useful to a future historian.[146] I found Crick's delightful autobiography altogether rather more "convincing" than Watson's account, although I deplore Crick's seeming indifference to the history of biochemistry, suggested by his statement that "enzymes are the machine tools of the living cell. They were first discovered in 1897 by Eduard Buchner, who received a Nobel Prize ten years later for his discovery."[147]

I hope that future historians will distinguish more sharply than has recently been the custom between the reception of the Watson-Crick model of DNA in 1953 and that of Watson's book in 1968. My own reaction to the two famous papers in *Nature* was one of cautious skepticism; when Sofia Simmonds and I were preparing the second edition of our *General Biochemistry* in 1957, we described the model as an "ingenious speculation."[148] The important idea of base pairing was consistent with the analytical data previously re-

145. Medawar (1968), p. 4.

146. Crick (1974, 1988). Perutz has noted his "long-held view that Watson invented the race of the Nobel Prize to make his book *The Double Helix* more exciting. By so doing, he generated the now widely held, deplorable misconception that this is what science is all about" (Perutz, 1989a, p. 8).

147. Crick (1988), p. 32.

148. Fruton and Simmonds (1958), p. 200. In this connection, I commend to the attention of future historians of molecular biology the article by Robert Sinsheimer (1957), based on a lecture he gave in November 1956.

ported by Erwin Chargaff, and the biological implications of this idea were obvious. I had learned during the 1930s, however, about the perils in the search for geometric simplicity in the structure of proteins and the uncertainties in the interpretation of the X-ray diffraction patterns given by fibrous macromolecules, and I was therefore loath to accept the model as the "solution" or "discovery" of a unique three-dimensional structure for chromosomal DNA. Although I may be classified by some biologists as a "crass biochemical reductionist," and have long considered biological systems to involve the operation of complex assemblies of their chemical constituents, as a chemist I was not ready to accept the view that what is aesthetically pleasing, or thermodynamically most stable, necessarily represents the only possible structure of these constituents as they exist and function in biological systems. Nor were these doubts about the Watson-Crick model fully resolved in 1958 after the brilliant experimental work of Matthew Meselson and Franklin Stahl in confirming the aspect of the Watson-Crick model which predicted "semi-conservative" replication of DNA strands, and of Arthur Kornberg and his associates in confirming the idea of base pairing in anti-parallel DNA strands.[149]

Among the more perceptive reviews in 1968 of Watson's book was the one by Robert Merton, who began as follows:

> This is a wonderfully candid self-portrait of the scientist as a young man in a hurry. Chattily written with pungent and ironic wit and yet with an almost clinical detachment, it provides for the scientist and the general reader alike a fascinating case history in the psychology and sociology of science as it describes the events that led up to one of the great biological discoveries of our time. I know of nothing quite like it in all the literature about scientists at work.[150]

I call attention to the last sentence in this excerpt because my own first reaction to the book was to compare it with a work of fiction which I read as a youth in 1927, and which impelled me to aspire to a career in biochemical research. That book was Sinclair Lewis's *Arrowsmith*, written with the help of the microbiologist Paul de

149. Meselson and Stahl (1958); Lehman et al. (1958).
150. Merton (1968b), p. 1. See also Chargaff (1968).

Kruif.[151] As a literary form, autobiography is, in many respects, comparable to the novel, and presents difficult problems in the separation of fact from fiction.[152] It must also be acknowledged, however, that on occasion a novel about scientists has provided valuable insights into the peculiarities of the scientific life in particular times and places. In addition to Lewis's *Arrowsmith*, the works of the novelist (and former physicist) Charles Snow may perhaps most readily come to mind, but I would also call attention to the former chemist Harry Hoff, who has written novels under the pen name of William Cooper.[153]

In *Arrowsmith* the hero is the immunologist Max Gottlieb, an unwordly German Jew. In exalting the virtues of the disinterested scientific life, the novel is in the tradition of romanticism; in describing the difficulty of pursuing such a life in the United States of the 1920s, the novel is also a work of social criticism. From my point of view as a fifteen-year-old Jewish boy, a recent immigrant from Poland about to enter Columbia College, it was the character of Max Gottlieb which provided the main inspiration. Since then I have reread the book many times, certainly more critically, but always with an appreciation of its morality. I have found this quality in many autobiographies by noted scientists, most recently in those of François Jacob and of Arthur Kornberg.[154] On my first reading of *The Double Helix*, its most striking feature was, for me, its amorality. The unabashed account of Watson's striving for success during the 1950s would have been, in my opinion, less deplorable if, in his maturity, he had acknowledged more generously that what Crick and he had achieved owed much to the published work of other scientists. It may be argued that, if he had done so, Watson would have lessened the dramatic quality of his story, and that he would have been a hypocrite to conceal his ignorance and his less than admirable activities in pursuit of renown. As one who has known many young American biochemists, I readily agree that their aspirations often were accompanied by the kind of behavior described by Watson. I would also agree that such behavior was not uncommon, but to a much lesser degree, in some areas of scientific inquiry before the Second World

151. Among the more recent writings about *Arrowsmith* are those by Schorer (1961), de Kruif (1962), Rosenberg (1976), pp. 123–131, and Lowy (1988).

152. See Lejeune (1989).

153. Snow (1934, 1960); Cooper (1953).

154. Jacob (1987); Kornberg (1989).

War. However, in recollection of the influence of *Arrowsmith* on my own youthful aspirations, I asked in 1968 what effect Watson's book might have on the values of young people entering the biochemical sciences at that time. That question has been raised by others.[155] I hope that future historians of these sciences will find, along with the "anti-heroes" of *The Double Helix*, some Gottliebs.

I have dwelt on *The Double Helix* because of the wide attention it has attracted, but many of the problems this book presents to the critical historian are also evident in other recent book-length autobiographies by noted biologists and chemists, books which I have read with admiration of the research achievements of the authors, but with variable response to the manner in which they have described their rise to eminence. A partial list of these authors includes Chargaff, Crick, du Vigneaud, Hastings, Jacob, Kornberg, Krebs, Lipmann, Luria, McCarty, Medawar, and Todd.[156] Shorter autobiographies are available for many investigators in the biochemical sciences in the series of volumes edited by Semenza[157] and in successive volumes of *Annual Reviews of Biochemistry*, as well as in other sets of the *Annual Reviews* series. Although the primary emphasis in most of these books and articles is on the past activities of the authors in research or administration, these writers also reveal, in Medawar's words, "truths about themselves." For example, in the autobiography of Alexander Todd, I found the passage relating to his declination of the invitation to become the successor of Hopkins in Cambridge to suggest something of his scientific style. Todd summarized his reactions as follows:

> (1) There was no real unity of purpose in the department. It was a series of little independent kingdoms sharing the departmental budget between them and the only gesture to unity was an almost sycophantic attitude to Hopkins on the part of the leaders in each of them.
> (2) With the exception of Robin Hill and F. G. Hopkins himself the staff seemed to have little or no interest in the only aspects of biochemistry in which I had any expertise.

155. See Chargaff (1974), Donohue (1976), and Davis (1980).

156. Chargaff (1978); Crick (1988); du Vigneaud (1952); Hastings (1989); Jacob (1987); Kornberg (1989); Krebs (1981); Lipmann (1971); Luria (1984); McCarty (1985); Medawar (1986); Todd (1983).

157. See Semenza (1981, 1982, 1983, 1986).

(3) The teaching courses were to my mind thoroughly inadequate as regards their chemical content and could produce no students who would fit into my type of work.
(4) I knew that I had a large number of people who wished to do research with me and on whom the progress of my work depended; there was simply no room to accommodate them in the Cambridge biochemical laboratories.[158]

Such statements suggest that the historian of the biochemical sciences, in addition to "listening at keyholes," can learn much about the behind-the-scenes features of the scientific enterprise, and about the personal relations of noted biologists or chemists, from the informed and critical study of their autobiographies.[159]

In recent years, professional historians and sociologists of science have shown increased interest in the ancient scholarly technique of prosopography, defined by Lawrence Stone as "the investigation of the common background characteristics of a group of actors in history by means of a collective study of their lives."[160] Although the purposes of such collective biographical studies have varied widely, with few exceptions they have dealt with groups of people belonging to "elites." As applied to the population of scientists, the term "elite" has generally come to connote a set of individuals who, by virtue of their election to such bodies as the Royal Society of London, stand out among their fellow scientists.[161] During the twentieth century, there has emerged what has been termed an "ultra-elite" of winners of Nobel Prizes in physics, chemistry, and physiology or medicine.[162] There is no doubt that the "Matthew effect" or "accumulation of advantage" has been operative in the enhancement of the social status of Nobel Prize winners in the biochemical sciences,[163] nor that (with perhaps a few exceptions) all the awards were applauded by informed members of the scientific community. For the historian who is interested in the development of the inter-

158. Todd (1983), p. 62.

159. Russell (1988a); this article contains a list of twenty-one autobiographies. See also Millar and Millar (1988) for autobiographies of chemists.

160. Stone (1971), p. 46.

161. See Shapin and Thackray (1974), Pyenson (1977), and Rehbock (1987). For the word "elite" see Williams (1985), pp. 96–98.

162. Zuckerman (1977), pp. 11–13.

163. Merton (1968a, 1988).

play of theory and practice in these sciences, however, the role of the Nobel Prizes in the advancement of knowledge is a matter which merits closer study. In particular, can the historian gain an understanding of that development solely through the collective biography of scientific elites, without taking into account the life and work of the "lesser folk" who contributed, however modestly, to the achievements of notable scientists, either as their predecessors or as their junior associates? Moreover, can the influence of a noted scientist on the later development of his or her area of inquiry be assessed solely in terms of the successes of the next generation of members of the "scientific elite"?

Professional historians of the chemical and biological sciences may rightly argue that, in view of their relatively small number and of the many problems concerning elites which still await study, the task of considering large scientific populations, with attention to the life and work of individual "ordinary" scientists, is ill-advised. As several historians have shown, however, the use of the technique of collective biography in the study of nineteenth-century research groups can provide valuable insights into important problems left untouched by concentration on noted individual scientists or on elite associations.[164] Admittedly, the search for biographical information about the "lesser folk" who did not attain leading positions at universities or research establishments can be a difficult one. Their names are usually absent from the major national biographies and only appear in the comprehensive bibliographies of scientific and medical publications during the nineteenth and twentieth centuries. Biographical dictionaries of scientists having a scope comparable to the successive editions of *American Men* (later *Men and Women*) *of Science* are not available for other countries. Consequently, the search for biographical information about "ordinary" scientists, although perhaps less challenging than that of Marc Bloch in his study of the role of artisans and peasants in feudal society,[165] requires some of the same assiduous and painstaking effort to find reliable data pertinent to the historical problem under investigation. To this I add the opinion that the use of such data for statistical analysis, while frequently providing a glimpse of historical

164. Morrell (1972); Geison (1978); Snelders (1984); Klosterman (1985); Servos (1990).

165. Bloch (1961).

trends, obscures what I believe to be one of the main purposes of scholarship in the history of the chemical and biological sciences: that is, to seek understanding of the ways in which individual scientists, however long or brief their research activity may have been, or whatever merit was attached to their contributions by contemporaries or by later scientists, participated in the development of knowledge in their areas of scientific inquiry.

CHAPTER FIVE

Reflections on the Biochemical Literature

After the attack on Charles Snow by F. R. Leavis,[1] Aldous Huxley took occasion to compare the languages of literary artists and of scientists:

> Like the man of letters, the scientist finds it necessary to "give a purer sense to the words of the tribe." But the purity of scientific language is not the same as the purity of literary language. The aim of the scientist is to say only one thing at a time, and to say it unambiguously and with the greatest possible clarity. To achieve this, he simplifies and jargonizes. In other words, he uses the vocabulary and syntax of common speech in such a way that each phrase is susceptible of only one interpretation; and when the vocabulary and syntax of common speech are too imprecise for his purposes, he invents a new technical language, or jargon, specifically designed to express the limited meaning with which he is professionally concerned. At its most perfectly pure, scientific language ceases to be a matter of words and turns into mathematics. . . . When the literary artist undertakes to give a purer sense of the words of his tribe, he does so with the express purpose of creating a language capable of conveying, not the single meaning of some particular science, but the multiple significance of human experience, on its most private as well as on its more public levels.

1. Snow (1959); Leavis (1962).

> He purifies, not by simplifying and jargonizing, but by deepening and extending, by enriching with allusive harmonies, with overtones of association and undertones of sonorous magic.[2]

As an example of the language of science, Huxley offers a sentence taken from "the highly purified languages of biochemistry, cytology and genetics":

> A special form of ribonucleic acid (called messenger RNA) carries the genetic message from the gene, which is located in the nucleus of the cell, to the surrounding cytoplasm, where many of the proteins are synthesized.[3]

Later in the same book, Huxley writes:

> Science is a matter of disinterested observation, unprejudiced insight and experimentation, patient ratiocination within some system of logically correlated concepts. In real-life conflicts between reason and passion the issue is uncertain. Passion and prejudice are always able to mobilize their forces more rapidly and press the attack with greater fury; but in the long run (and often, of course, too late) enlightened self-interest may rouse itself, launch a counterattack and win the day for reason.[4]

I have quoted these passages from Huxley's book, written shortly before his death, not only because he was a great literary artist but also because his acquaintance with contemporary developments in the biochemical sciences reminds one that he was a member of a family which included many noted biologists. These views of a distinguished man of letters raise questions about the writing and publication of research articles in the biochemical sciences, the development of the vocabulary used in such articles, and the social role of these articles in the relationships among members of scientific "tribes" with one another and to their patrons.

2. Huxley (1963), pp. 12–13. (Quoted by permission of the estate of Aldous Huxley.)
3. Ibid., p. 13.
4. Ibid., pp. 68–69.

"The Words of the Tribe"

Since the eighteenth century, there has been a continuous effort to give a "purer sense" to the words and symbols used to denote the objects and processes encountered during the interplay of biology and chemistry. Throughout the course of that effort, the languages have changed in response to new empirical discovery, to the introduction of new methods, and, above all, to transitions in the conceptual frameworks of biological and chemical thought. In an apt aphorism, the philosopher Otto Neurath likened languages to a ship which is being rebuilt at sea while keeping it afloat. At any time, words inherited from the past to denote particular natural objects or classes of such objects have been retained, but the meaning of such words has undergone considerable modification; examples to be discussed later in this chapter are Lavoisier's "oxygène," Mulder's "protein," and Kühne's "enzyme." Likewise, terms used to denote particular relations among natural objects, for example "affinity," survived, but acquired new meanings in the various contexts in which such terms were used. The technical languages, or "jargons,"[5] contained not only neologisms fashioned out of Greek or Latin roots but also words drawn from contemporary national languages and given special meanings, as in the current usage by molecular biologists of the word "messenger" for a biochemical object and the word "recognition" for a relation between biochemical objects. Such usage, based on analogy, has in the past been a source of ambiguity and controversy, but as the philosopher W. V. O. Quine stated, "Devices conceived in error have had survival value, and are to be assessed on present utility. But we stand to increase our gains by clearing away confusions that continue to surround them; for clarity is more fruitful on the average than confusion, even though the fruits of neither are to be despised."[6]

These considerations are relevant to the problems encountered in the classification and nomenclature of surviving and extinct biological organisms, in the assignment of names given to the parts of organized biological systems and the processes in such systems, in the successive reforms of the nomenclature of chemical substances, and in the transfer of words derived from biology or chemistry to

5. See Williams (1985), pp. 174–176, for a discussion of this word.
6. Quine (1960), p. 123.

the other discipline. These problems have not been, as some biologists, chemists, or biochemists may have believed, trivial concerns of pedantic editors of scientific journals, but have reflected the state of the development of an area of inquiry at a particular time. I believe, therefore, that an appreciation of the changes in the language used within a research discipline is essential for the understanding of the development of the conceptual framework of that discipline.

From the time of Carl von Linné and Georges Louis Buffon, the classification and nomenclature of plant and animal species have undergone extensive revision. Many noted biologists, including Charles Darwin, participated in the effort to devise a system based on some "natural order" (variously defined as "God-ordained," "rational," or "of common descent"). The recognition of the limitations of the "two-kingdom" system, especially in relation to such classes of organisms as algae, bacteria, fungi, and protozoa, led to more complex systems of classification.[7] In the recent past the effort has involved, in addition to the traditional grouping of organisms according to their outward appearance and prominent internal features (now termed "phenetics"), classification based on genealogy (now termed "cladistics") or on both genealogy and evolutionary analysis. Since the number of names assigned to known surviving and fossil organisms is now in the millions, the task of modern taxonomists is both difficult and uncertain.[8] The Linnaean binary Latin nomenclature of biological species was the result of an effort to group known organisms according to a particular conceptual system, and changes in that nomenclature have reflected changes in the concepts which have guided the formulation of successive schemes of biological classification. Of necessity, to facilitate communication among biologists, international codes of nomenclature in zoology, botany, and bacteriology had to be established and repeatedly revised.[9] The value of an arbitrary assignment of a preferred name to an organism previously given different names by various investigators is obvious, especially in view of the practice of honoring the presumed discoverer (as in *Escherichia coli*, named by Theodor Escherich *Bacillus coli communis*). Such definitions have been

7. See, for example, Whittaker (1969) and Woese et al. (1990).
8. Mayr (1968, 1982, 1989).
9. A useful guide to these international codes was prepared by Jeffrey (1973).

needed, for example, by investigators in the biochemical sciences who used selected (often uncommon) organisms especially suited for the study of particular physiological processes.[10] In addition, the use of a classical language helps to reduce the problems involved in translating the vernacular names of a biological species from one modern national language to another.

In the nomenclature of the parts of biological organisms, it has been a matter not only of the names given to discrete internal structures made visible by dissection or by microscopy, but also of the names assigned to materials obtained from such organisms by the ancient chemical methods of "extraction," followed by evaporation of the solvent or precipitation of solid material, or of "distillation" of the more volatile products derived by the application of heat to materials from biological sources.[11] During the eighteenth century, it became customary to name such chemical products according to their biological source. For example, in 1749 Andreas Sigismund Marggraf obtained from red ants *(Formica rufa)* a volatile product to which the French term "acide des fourmis" was assigned; in 1791, it was renamed "acide formique." On the other hand, Carl Wilhelm Scheele eschewed dry distillation and preferred extraction methods, followed by precipitation of sparingly-soluble salts which were then carefully converted to the parent acids by means of "oil of vitriol" (sulfuric acid). In this manner he isolated products from wood sorrel *(Oxalis acetosella)*, lemons, apples, and milk; the French names given these products were "acide oxalique," "acide citrique," "acide malique," and "acide lactique." Although these terms, derived from Latin roots, were translated into the corresponding English equivalents, German chemists preferred "Ameisensaüre," "Zitronensaüre," "Aepfelsaüre," or "Milchsaüre," and Russian chemists followed the German example, as in "limonnaya kislota." These differences in the "jargon" of nineteenth-century chemists became less troublesome as a consequence of the development between 1810 and 1840 of increasingly accurate methods for the determination of the elementary composition of well-defined chemical substances such as oxalic, citric, malic, or lactic acid, and by the subsequent acceptance of a uniform symbolism to denote their elementary composition and the arrangement of their atoms. Thus, with the acceptance of relative

10. Schmidt-Nielsen (1967).
11. Holmes (1971).

atomic weights for carbon (C = 12), oxygen (O = 16), and hydrogen (H = 1), the "empirical" formula of, say, oxalic acid was taken to be $C_2H_2O_4$ and its "rational" formula could be written as HOOC-COOH, no matter what words French, English, German, or Russian chemists used to name it.

The transformation of the language used to denote chemical substances was, as in the Linnaean nomenclature of biological organisms, a consequence of efforts to devise a coherent system of classification of such substances. During the eighteenth century, Torbern Bergman, Louis Bernard Guyton de Morveau, and Antoine Lavoisier, in part inspired by the writings of the Abbé Bonnot de Condillac, initiated an effort to give "a purer sense of the words of the tribe."[12] Among the early fruits of that effort was a reform of the nomenclature of substances considered to be chemical elements, notably the constituents of common air later called oxygen and nitrogen. Both quantitative measurement and biological experimentation played significant roles in that development, and Frederic Holmes has documented in detail the close connection of Lavoisier's researches on the combustion of charcoal and on the respiration of animals.[13] What Lavoisier called in 1777 "l'air emminément respirable" became "principe acidificant ou principe oxygine" and later "oxygène," while the non-respirable portion of common air was first called "mofette atmospherique" and then "azote" (Greek, *azoe*, not life). The subsequent term "nitrogen", introduced after the naming of "hydrogen," was not adopted by many nineteenth-century French chemists, who used the symbol "Az" in place of "N" in writing empirical formulas. The German chemists also went their own way, and still refer to oxygen, hydrogen, and nitrogen as *Sauerstoff, Wasserstoff,* and *Stickstoff,* respectively.

The effort initiated by Bergman and by Lavoisier's group was continued during the nineteenth century by many chemists, including such notables as John Dalton, Jöns Jacob Berzelius, Auguste Laurent, August Kekulé, and Dmitri Ivanovich Mendeleev, and reflected the successive transitions in the meaning of such chemical concepts as element, compound, atom, equivalent, and molecule.[14]

12. For Condillac, see Gillispie (1971).

13. Holmes (1985).

14. Crosland (1962), pp. 133–214; Flood (1963); Hogben (1969), pp. 28–37; Dagognet (1969); Eklund (1975); Rocke (1984); Anderson (1984); Alborn (1990); Dear (1991).

These transitions, attended by much controversy, were the outcome of new empirical knowledge drawn from many sources—crystallography, quantitative analysis, measurement of such physical properties as vapor density or specific heat, and the intensive study of a large variety of chemical reactions.[15]

In the famous 1787 reform of chemical nomenclature proposed by Morveau, Lavoisier, Berthollet, and Fourcroy, a consistent system was devised for the naming of acids, oxides, and salts of mineral origin; for example, the alchemical name "sucre de Saturne" was replaced by "acetate de plomb." However, many of the compounds obtained from animal or plant sources presented difficulties which were not resolved. Among these "organic" substances were products then named (in English) sugar, starch, gluten, albumen, fixed oils, and volatile oils. During the first decades of the nineteenth century, pharmacists and physicians in France, Germany, and Britain isolated many more substances of this kind, and new names, usually ending in *-ine* (or *-in*) were invented. This development was noticed outside the chemical community. For example, in his novel *La Peau de Chagrin*, published in 1831, Honoré de Balzac recounts a conversation with the "famous chemist" Japhet:

> "Well, my old friend," said Planchette upon seeing Japhet seated in an armchair and examining a precipitate, "how goes it in chemistry?"
>
> "It is asleep. Nothing new. The *Académie* has in the meantime recognized the existence of salicine. But salicine, asparagine, vauqueline, digitaline are not new discoveries."
>
> "If one is unable to produce new things," said Raphael, "it seems that you are reduced to inventing new names."
>
> "That is indeed true, young man!"[16]

During the succeeding century and a half, countless new names were invented for products obtained from biological sources, but a large proportion of these names disappeared from successive organic-chemical or biochemical treatises, either because the products were later found to be mixtures and the "pure" compounds given other names, or because new systematic chemical classifications

15. See Freund (1904).
16. My translation from Balzac (1925), p. 239.

suggested the replacement of older terms. In organic chemistry, the first steps in devising an international code were taken during the 1890s and led to the so-called Geneva system, which was followed by revisions during the 1930s and 1950s; these efforts, under the auspices of the International Union of Pure and Applied Chemistry, have continued to the present.[17]

The successive systems of chemical classification, based on rules for designating the structure of classes of compounds (as in the case of the alcohols R-OH, where R may be methyl, ethyl, and so on), have clarified many of the problems encountered in the nomenclature of chemical substances of biological interest. During the course of the emergence and growth of the biochemical sciences, however, there has been an alternative basis of classification, namely in terms of the demonstrated (or presumed) functional role, in living organisms, of a particular class of chemical substances. For example, the word "hormone" was introduced in 1905 by Ernest Henry Starling to denote a class of "chemical messengers" which act in the animal organism as regulators of some physiological processes independently of the nervous system.[18] At that time, the chemical structure of only one of the members of this class (adrenalin or epinephrine) had been established, but in succeeding decades the structures of the active hormonal principles of the thyroid, pancreas, adrenal gland, pituitary, sex organs, and gastrointestinal tract were elucidated, and shown to belong to very different kinds of organic compounds (steroids, peptides). Likewise, the word "vitamine" (later shortened to "vitamin") was proposed by Casimir Funk in 1912 to denote the class of substances which were required in the diet of some animal species, and which Frederick Gowland Hopkins had previously called "accessory food factors."[19] Subsequent chemical work demonstrated that this class of nutrients includes organic compounds of widely different chemical nature (derivatives of pyridine, thiazole, thiophene, glucose, cholesterol, iso-alloxazine, carotene, tetrapyrroles, and so forth). The disparity between the classifications based on chemical structure and those based on physiological function has repeatedly led to the introduction of neologisms which, if

17. See Hurd (1961), Cahn and Dermer (1979), and Verkade (1985).

18. Starling (1905); see also Wright (1978) and Simmer (1978).

19. Funk's fame derives principally from his neologism and his entrepreneurship. His experimental contributions to the study of vitamins were far less distinguished than those of his contemporaries. See Harrow (1955) and Ostrowski (1986).

widely accepted, were sources of ambiguity and controversy. Indeed, in the biochemical sciences, the task of giving "a purer sense to the words of the tribe" still presents challenges, especially in the case of such terms now in common usage as "gene," "protein," or "enzyme." With the growth of knowledge about the chemical structure and physiological function of the classes of entities to which such names have been assigned, these terms have acquired meanings rather different from those intended by their originators, and the changes in meaning have reflected transitions in the conceptual framework of the biochemical sciences. During the course of these transitions, what has frequently been at issue is the relationship between the chemical nature and the biological function of an entity to which a particular name was given.

I have mentioned in a previous chapter some aspects of this problem as it relates to the word "gene," introduced in 1909 by the botanist Wilhelm Ludvig Johannsen. This word is now applied to a class of chemical entities—oligonucleotide segments of chromosomal DNA—and a multiplicity of entities defined in terms of cytological, cross-breeding, mutational, or physiological studies on biological organisms.[20] The word "protein" is of more ancient vintage, and was put forward (at the suggestion of Berzelius) by the agricultural chemist Gerrit Jan Mulder in 1838, to denote "the organic oxide which is the base of fibrin and albumin" and to emphasize the seemingly ubiquitous presence in biological systems and the nutritional importance of the class of substances to which they belong.[21] Mulder's word was largely rejected during the succeeding years of the nineteenth century, and both chemists and biologists preferred "Eiweissstoff," "matière albuminoide," or "albuminous body" (later "proteid"). The word "protein" gained wide acceptance only after 1900, in large part because of its adoption by the noted organic chemist Emil Fischer in his program of research on the amino acid composition of proteins and on the synthesis of peptides. It is, I believe, fair to say that thus far all the many attempts to devise a systematic classification of proteins have failed to gain wide acceptance, whether such attempts were based on the similarity in such chemical properties as solubility in various solvents or (later) details of their three-dimensional structure, or on similarity in their

20. Stadler (1954); Carlson (1966); Burian (1985).
21. See Fruton (1972), pp. 95–101.

biological function as enzymes, immune bodies, contractile fibers, or carriers of oxygen.[22] Indeed, the extensive knowledge gained since the 1960s about the variations in the amino acid sequences of proteins suggests that the problem of their classification is, in some respects, not unlike that of the taxonomy of biological species, except for the fact that an individual organism produces many proteins. As I noted previously, the amino acid sequences of proteins have been added to the phenotypic characters which may be used for evolutionary analysis and biological classification. Perhaps a systematic chemical classification, based solely on the apparently "invariant" segments of related proteins, may be deemed in the future to be useful for their nomenclature.

Such a chemical classification appears to have been possible for the group of enzyme proteins which catalyze the hydrolysis of amide (CO-NH) and ester (CO-OR) bonds. The extensive recent chemical studies on the mechanisms of the catalytic action of many members of this group of enzymes, together with studies on their amino acid sequences and three-dimensional structures, have led to the recognition that their "catalytic sites" are, in general, of four kinds: those in which catalysis depends on the participation of the hydroxyl group of a particular serine unit, or of the sulfhydryl group of a particular cysteinyl unit, or of the carboxyl group of a particular aspartic acid unit, or of a metal ion such as zinc.[23] The action of these groups of enzymes involves, in all cases, the participation of the side-chain groups of other amino acids brought into proximity to these essential units by the specific folding of the polypeptide chain of the enzyme protein. The resulting "catalytic sites" are therefore characteristic of the members of each of the four groups, and in the case of the enzymes which catalyze the cleavage of the peptide bonds of proteins and peptides it has become customary to refer to "serine proteinases," "cysteine proteinases," "aspartic proteinases," and "metal proteinases."

Such classifications, although widely accepted among investigators in an area of biochemical inquiry, have not received endorsement in the reports of the successive international commissions

22. For a view of the status of protein chemistry during the first decade of this century, see Mann (1906). For later discussion of the classification of proteins, see Needham (1965), Richardson (1981), and Fantini (1984).

23. See Neuberger and Brocklehurst (1987).

which have attempted to devise a rational system of classification and nomenclature of enzymes, based on the chemical reactions which they catalyze; the latest of these reports (1984) lists 2,477 enzymes.[24] I believe it is fair to say that, in contrast to the systems of the classification of biological organisms or of organic-chemical compounds, the reports on enzymes have not gained general acceptance among workers in the biochemical sciences. In part, this response is a consequence of the attempt to superimpose on a rational chemical classification the consideration of the demonstrated (or presumed) physiological function of an enzyme. Thus, although it is generally agreed that the chemical role of an enzyme is to catalyze the attainment of equilibrium in a chemical reaction, the latest report continues to base the "recommended name on either direction of reaction, and is often based on the presumed physiological function."[25] The ambiguity introduced by such nomenclature is evident in the appearance of articles about "ATP-synthase (H^+-ATPase)."[26] According to the 1984 report, the suffix *-ase* after the name of the substrate denotes a "hydrolase," whereas what had been called synthases are included in a different class, termed "lyases." Moreover, although it is acknowledged in the report that

> a number of hydrolases acting on ester, glycosyl, peptide, amide or other bonds are known to catalyse not only hydrolytic removal of a particular group from their substrates, but likewise the transfer of this group to suitable acceptor molecules. In principle, all hydrolytic enzymes might be classified as transferases, since hydrolysis itself can be regarded as transfer of a specific group to water as the acceptor. Yet, in most cases, the reaction with water as the acceptor was discovered earlier and is considered as the main physiological function of the enzyme. This is why such enzymes are classified as hydrolases rather than transferases.[27]

24. Webb (1984). The first of these reports, published in 1961, was largely based on the scheme proposed by Dixon and Webb (1958). An earlier proposal for the reform of the classification of enzymes had been offered by Hoffmann-Ostenhof (1953); see also Fruton and Simmonds (1958), pp. 215–216, 273–274.

25. Webb (1984), p. 5.

26. See, for example, Futai et al. (1989).

27. Webb (1984), p. 9.

The framers of this report thus restricted the chemical concept of "group transfer" only to those enzymes which they chose to call "transferases," and decided to retain, on historical and physiological grounds, the outworn definition of "hydrolases." The confusion generated by attempts to combine chemical and biological definitions in a classificatory scheme has had its echoes, as in the article by John Maddox, for whom "'ribonuclease' after all, is the name for a biochemical function and not a chemical structure."[28] The biochemists—Kunitz, Anfinsen, Stein, Moore, Richards, Hofmann—who worked on the chemistry of pancreatic ribonuclease A were apparently mistaken in their belief that the crystalline material they handled was a chemical substance of defined structure! Moreover, when these men worked on this enzyme, its physiological function was a subject of some discussion, because it was then known that the mammalian organism can synthesize the purines and pyrimidines of RNA from smaller metabolites, and does not depend on the action of the ribonuclease in the pancreatic secretion into the intestine for a supply of the structural units of the ribonucleic acids of the body tissues. I must add to my criticism of some aspects of the 1984 report my agreement with its disapproval of the indiscriminate use of the suffix *-ase* to denote supposed catalytic agents such "permease," "translocase," and "replicase." However, a measure of the influence of the report is provided by the fact that such terms (for example, "primase," "resolvase," "transposase") have continued to proliferate in the biochemical literature.

I have dwelt on the problem of the classification of enzymes in part because of its relevance to some of the work during the 1950s of my biochemical research group, and because as a member of one of the international commissions on biochemical nomenclature during the 1960s, I failed to convince my colleagues of what I believed to be the error of their ways. For the purposes of this book, a more important reason for dwelling on the present state of this problem is that it reflects a continuing tension in the interplay of biological and chemical thought. That tension was already evident during the 1870s, in the reception accorded to Willy Kühne's introduction of the word "Enzyme" into biochemical language. His stated aim was to emphasize the biological character of such physiological processes as oxidations and fermentations, and to distinguish such

28. Maddox (1976).

processes from the hydrolytic action of agents ("Enzyme") which had been extracted from biological sources, and had been termed "soluble ferments" or "unorganized ferments." This attempt to "give a purer sense" to the German word "Ferment" was not greeted cordially by the more chemically-minded physiological chemist Felix Hoppe-Seyler. I have discussed elsewhere some of the history of the eventual general adoption of the term "enzyme," with a meaning rather closer to Hoppe-Seyler's definition of "Ferment," in place of that German term. Leading British biochemists adopted the newer term more readily than did the Germans or the French; the latter clung to the word "diastase."[29]

During the twentieth century, some of the greatest confusion in the terminology of enzymes arose from the writings of Otto Warburg, one of the greatest biochemists of his time. The universal admiration for Warburg's brilliant experimental discoveries, especially during the 1930s, demanded attention to the words he introduced, with seeming caprice, to denote new biochemical entities. This practice was not only an expression of Warburg's self-esteem, but also a reflection of his earlier rejection of the evidence for the existence of soluble "oxidases" and "dehydrogenases" and for their role in the intracellular oxidation of metabolites by molecular oxygen. For example, in 1929 Warburg wrote:

> If one calls oxidases ferments which transport molecular oxygen, then the extracts contain oxidases, and if one classifies the oxidases, as is customary in ferment chemistry, according to the observed actions, then one has in the extracts different oxidases, glucose oxidase, alcohol oxidase, indophenol oxidase, and so on. Strictly speaking there are as many different oxidases as extraction experiments. If the extract-oxidases had been preformed in the cell, a single type of cell would contain innumerable oxidases. But the multiplicity of oxidases in the living cell would be in opposition to a sovereign principle in the living substance. . . . Therefore, if many different oxidases have been found in extracts of a cell type, these were not ferments which were already present in the living cell, but are rather products

29. Fruton (1990), pp. 83–90. Plantefol (1968) has provided a useful summary of the change in the attitude of French biochemists toward the word "enzyme." See also Teich (1981).

of the transformation and decomposition of a single homogeneous substance present in life.[30]

This statement, apart from indicating Warburg's adherence, before 1930, to a "protoplasmic" theory of life, was also a claim that the "Atmungsferment" he had identified as the agent of intracellular respiration was not related to other iron-porphyrin (heme) "oxidases," notably the cellular components which David Keilin had named "cytochromes."[31]

Warburg's remarkable researches of the 1930s on enzymes (which he termed "Fermente") began with the study of the oxidation of glucose-6-phosphate by molecular oxygen in the presence of disrupted mammalian erythrocytes.[32] During the course of this work, he and Walter Christian identified an entity, which Warburg named "das gelbe Ferment," composed of a protein material and a yellow pigment. The pigment could be dissociated from the protein and was found to undergo reversible oxidation and reduction, and the reduced form could be oxidized by molecular oxygen. Hugo Theorell showed the pigment to be the 5′-phospho derivative of riboflavin (vitamin B_2), an iso-alloxazine compound, and Warburg considered the flavin derivative to be the catalytically active group of the "gelbe Ferment." This view was endorsed by the noted organic chemist Richard Kuhn, a pupil of Richard Willstätter: "R. Willstätter thought that an enzyme consists of a colloidal support and an active group. The explanation given by O. Warburg and H. Theorell for the structure of the yellow enzyme illustrates exactly this conception."[33] Later during the 1930s other "yellow enzymes" were discovered; in some of them the pigment was found to be flavin adenine dinucleotide (FAD) in place of riboflavin phosphate, which came to called flavin mononucleotide (FMN). Warburg referred to this group of "flavoprotein enzymes" as "Alloxazin-proteide."

For the oxidation of glucose-6-phosphate by the flavoprotein to proceed, another protein fraction was needed, and Warburg called it "Zwischenferment" rather than "glucose-6-phosphate dehydrogenase." There was also a requirement for a low-molecular-weight entity, and in a brilliant chemical investigation, Warburg's group

30. Warburg (1929), pp. 1–2.
31. Keilin (1966), pp. 194–203.
32. See Fruton (1972), pp. 301–305, 334–339, 353–358.
33. Kuhn (1935), p. 921.

showed the substance from erythrocytes to be a derivative of nicotinic acid amide (3-pyridinecarboxylic acid amide), and he called the compound triphosphopyridine nucleotide (TPN). It was immediately recognized that this compound was related to the co-ferment of alcoholic fermentation, discovered in 1906 by Arthur Harden and William John Young and named by them "cozymase." Warburg's group then showed "cozymase" to be the corresponding diphospho compound, which acquired the name "coenzyme I" and the abbreviations CoI or DPN, and TPN was also termed "coenzyme II" or CoII; at present, the preferred names are nicotinamide adenine dinucleotide (NAD) and its phospho derivative (NADP), respectively.

As in the case of the "yellow enzymes," Warburg considered the pyridine nucleotides to be the catalytically active groups of the enzymes he declined to call dehydrogenases. This view was questioned in 1937 by members of Hopkins's Biochemical Laboratory in Cambridge:

> The question of how the coenzyme functions in the catalytic system has now become a matter of great dispute. The concept of "Zwischenferment" introduced by Warburg implies that the coenzyme combines with the dehydrogenase to form the catalytically active complex. What is ordinarily defined as a dehydrogenase is considered by Warburg to be merely a highly specific protein with no catalytic properties apart from its prosthetic group—the coenzyme. Euler and his school have accepted this view but they prefer to call the active complex the "holodehydrase."
>
> There is a good deal of evidence in favour of the view that the dehydrogenase is the seat of catalytic activity. . . . The classical definition of the dehydrogenase as the actual activating mechanism seems to be in fair agreement with the facts. No doubt the coenzyme combines with the dehydrogenase in the same way that the substrate does. But the function of the coenzyme seems to be that of a highly specific hydrogen acceptor which cannot be replaced by any other substance.[34]

Such opinions did not sway Warburg, for in subsequent reports on the dehydrogenases involved in the conversion of glucose to alcohol

34. Green et al. (1937), p. 948.

or lactic acid, he termed the agent (known to others as glyceraldehyde phosphate dehydrogenase) for the conversion of glyceraldehyde-3-phosphate to 3-phosphoglyceric acid "das oxydierende Gärungsferment," and he gave the name "reduzirende Gärungsfermente" to the agents for the conversion of acetaldehyde to alcohol (known as alcohol dehydrogenase) and of pyruvic acid to lactic acid (known as lactic dehydrogenase). Indeed, in a book published in 1948, containing some of the papers from his laboratory from 1932 until 1945, Warburg wrote in the preface that "the hydrogen-transferring ferments are the first ferments whose active groups were isolated."[35] By that time it was clear to most enzymologists that the dehydrogenases, like other highly purified enzyme proteins such as pepsin, possess "active sites" involving particular amino acid units; for example, in 1938 Louis Rapkine had provided evidence for the role of a cysteinyl unit of glyceraldehyde-3-phosphate dehydrogenase in the catalytic activity of this enzyme.[36] It was also widely recognized that advances in the knowledge about such active sites depended on progress in the determination of the amino acid sequences and three-dimensional structure of enzyme proteins.

I have devoted much attention to Warburg's terminology because, apart from his stubborn idiosyncrasies, it illustrates the difficulties presented by attempts to combine, in the formulation of a rational classification and nomenclature of biochemical substances, two kinds of properties. The chemical nature and catalytic action of individual enzymes can provide a sound basis for a start in their rational classification, but such a classification is, I believe, incompatible with one based on their demonstrated (or presumed) physiological function. For Warburg, in the process of the biological oxidation of glucose-6-phosphate by molecular oxygen, what we now would term a dehydrogenase-NADP unit or an entire flavoprotein is indeed a catalyst in the overall process, for the NADP or FMN system is continuously shuttling back and forth between its oxidized and reduced states. Moreover, his choice of terms indicated the direction of a particular chemical reaction in a particular physiological pathway. From the point of view of a chemically-minded biochemist of the late 1930s, however, NADP or FMN is one of the two substrates in a bimolecular reaction catalyzed by an active site

35. Warburg (1948), p. 7. See also Warburg (1938).
36. Rapkine (1938).

of unknown constitution in an enzyme protein, and, for the purposes of systematic classification, data on the acceleration of the attainment of equilibria in such individual oxidation-reduction reactions seemed to be more useful than those based on presumed physiological function in a biological organism. Indeed, with the development after about 1950 of new methods for the study of protein structure and enzyme action, the approach of chemically-minded enzymologists yielded valuable knowledge about the structure of the active sites of individual enzymes and the organic-chemical mechanisms operative in their catalytic action, as well as about the equilibria and rates of individual enzyme-catalyzed reactions. Although such "bio-organic" or "biophysical" experimentation could not, in itself, elucidate the physiological role of a particular enzyme, the chemical knowledge so provided has been indispensable for further study of its function in an organized biological system. In my opinion, this chemical knowledge is likely to be more useful for a rational classification of enzymes than the continued attachment of some biologically-minded biochemists to what "is considered to be the main physiological function of the enzyme."

It is perhaps not too far-fetched to suggest that the recent problems in the classification and nomenclature of biochemical substances are not dissimilar from some of those faced by the French reformers of chemical nomenclature in the last decades of the eighteenth century. Thus the terms "l'air vital" and "l'air emminnément respirable," with their physiological connotation, were abandoned in favor of a purely chemical designation.[37] If the word Lavoisier chose expressed the idea that "oxygène" is an acid-forming substance, the later development of chemistry confirmed Guyton de Morveau's view that "one should prefer a name which expresses nothing to a name which could express a false idea."[38] Indeed, as Lavoisier wrote, "we cannot improve the language of any science without at the same time improving the science itself; nor can we, on the other hand, improve a science without improving the language or nomenclature."[39] If the word "oxygène" later acquired a chemical meaning different from that assigned to it by Lavoisier, so do the words "protein" or "enzyme" have meanings rather different from those

37. Holmes (1985), pp. 317–318.

38. Guyton de Morveau (1782), pp. 374–375. For Guyton de Morveau, see Bouchard (1938) and Smeaton (1972).

39. Lavoisier (1790), p. xv.

intended by Mulder or Kühne. In the case of oxygen, the physiological connotation disappeared from chemical nomenclature, and the ambiguities in the meaning of the words now used to classify and name macromolecular constituents of biological systems suggest the need for further "improvement" of the languages of the biochemical sciences.

Many of the ambiguities in the classification and nomenclature of chemical substances according to their known constitution had been resolved by the end of the nineteenth century, but the problems posed by studies on the extent and rate of their interaction presented greater difficulty. After 1718, when Etienne François Geoffroy offered a scheme for classifying chemical substances according to their relative tendency *(rapport)* to form new products (as in the reaction of mineral acids with alkalis to form salts), the concept of "affinity" came to occupy a significant place in chemical thought. A prominent feature of this development was the application of Newtonian ideas of gravitational attraction to the interaction of chemical materials.[40] In particular, the Swedish chemist Torbern Bergman developed a system of chemical classification based on what came to be called (in English) "elective attraction," and the poet Goethe, a seeker of unity in nature, brought this chemical concept to public attention through his novel *Die Wahlverwandtschaften*, published in 1809.[41] During the first two decades of the nineteenth century, the concept was adopted in Germany and Britain by some philosophers and literary artists who sought such unity in the forces of nature, and some scientists (for example, Johann Wilhelm Ritter, Hans Christian Oersted, Humphry Davy) were influenced in their philosophical thought by these writers. For the development of chemical thought, however, the importance of the contributions of Bergman and especially of Lavoisier lay not only in their chemical ideas, but also in the improvements they made in the quantitative analysis of chemical substances, thus helping to bring mathematics into the language of chemistry.[42] In addition, the term "affinity" began to assume new meanings. In 1799, Claude Louis Berthollet demonstrated that the results of a chemical reaction between two substances depend not only on their affinity, but also on their relative quantities. This

40. Smeaton (1963).
41. For Bergman, see Smeaton (1970); for Goethe's book, see Adler (1990).
42. Levere (1971); Guerlac (1976); Holmes (1985); Lundgren (1990); Whitt (1990).

aspect of the problem did not receive much attention until the 1860s, with the reports by Marcelin Berthelot and Léon Péan de Saint-Gilles and by Cato Maximilan Guldberg and Peter Waage on what has come to be called the "law of mass action." The first paper (1864) by Guldberg and Waage began as follows:

> The theories which have been put forward thus far in chemistry in relation to the mode of the action of chemical forces are recognized by all chemists to be unsatisfactory. This applies to both the electrochemical and thermochemical theories; and it is doubtful whether it will be possible to find, from the development of electricity or heat which accompanies chemical processes, the laws according to which chemical forces act.[43]

The "mass action" principle brought into modern chemical language the word "equilibrium," inherited from ancient mechanics, which was to figure prominently in the later development of chemical thermodynamics.

In the passage just quoted, the "electrochemical theory" stemmed principally from the achievements of Michael Faraday, pupil and successor of Davy, whose experiments led him to consider chemical affinity to be an electrical phenomenon. A quotation from Hermann Helmholtz's memorable Faraday Lecture in 1881 is appropriate here:

> Now that the mathematical interpretation of Faraday's conceptions regarding the nature of the electric and magnetic forces has been given by Clerk Maxwell, we see how great a degree of exactness and precision was hidden behind the words, which to Faraday's contemporaries appeared either vague or obscure; and it is in the highest degree astonishing to see what a large number of general theorems, the methodical deduction of which requires the highest power of mathematical analysis, he found by a kind of intuition, with the security of instinct, without the help of a single mathematical formula. I have no intention of blaming his contemporaries, for I confess that many times I have myself sat hopelessly looking upon some paragraph of

43. See the German translation of the papers by Guldberg and Waage (1899), edited by Abegg; the citation is on p. 3. For an account of the development of the concept of chemical affinity, see Partington (1964), pp. 569–607.

> Faraday's descriptions of lines of force, or of the galvanic current being an axis of power, &c.[44]

Later in the same lecture, Helmholtz noted that Faraday often expressed "his conviction that the forces named chemical affinity and electricity are one and the same," and added that

> I think the facts leave no doubt that the very mightiest among the chemical forces are of electrical origin. The atoms cling to their electrical charges, and opposite electric charges cling to each other. If we conclude from the facts that every unit of affinity is charged with one equivalent either of positive or of negative electricity, they can form compounds. . . . You see that this ought to produce compounds in which every unit of affinity of every atom is connected with one and only one other unit of another atom. This, as you will see immediately, is the modern chemical theory of quantivalence, comprising all the saturated compounds.[45]

Helmholtz's lecture was delivered at a time when one of the major subjects of discussion among organic chemists was the question whether an atom could have a variable valence. Although Kekulé's idea of a fixed valence explained, as Helmholtz stated, the constitution and behavior of saturated carbon compounds, the growing body of empirical knowledge about other substances, particularly those to those to which Kekulé assigned "double bonds," invited speculation about "residual affinity." With the emergence of the electronic theory of valence during the second decade of the twentieth century, this term lapsed into disuse.[46]

The thermochemical approach to the definition of chemical affinity was advocated during the 1850s by Julius Thomsen, who considered the heat evolved in a chemical reaction (as measured in a calorimeter) to be a measure of affinity. With the introduction of the concepts of "entropy" and "free energy" (other terms were also used to denote these functions), and the subsequent development of chemical thermodynamics, the shortcomings of Thomsen's ap-

44. Helmholtz (1881), pp. 277–278.
45. Ibid., pp. 302–303.
46. See Russell (1971).

proach were widely recognized in the new physical chemistry which emerged during the 1880s.[47] For Wilhelm Ostwald, the principal leader in the academic establishment of this discipline, and until about 1909 a confirmed anti-atomist who exalted energy over matter, the concept of chemical affinity was especially attractive.[48] The journal he founded in 1887 bore the name *Zeitschrift für physikalische Chemie, Stöchiometrie und Verwandtschaftslehre.*

In more recent times, the concepts once represented by the term "chemical affinity" have been expressed either in the mathematical language of physical chemical theory or in words that denote the specific interaction of molecules (or regions of molecules) having "complementary" structures, as in Emil Fischer's famous lock-and-key analogy of enzyme-substrate combination. It is in the latter sense that biochemists now speak of "affinity chromatography" or the "affinity" labeling of the active sites of enzymes. Since the emergence of molecular biology, the related concept denoted by the word "specificity" has also acquired other meanings, such as that associated with the word "recognition," or with other words from ordinary language used in non-mathematical treatments of information theory.

The word "affinity" also appeared in the writings of nineteenth-century biologists who sought to devise "natural" systems of the classification of animals on the basis of their resemblances in form and function. For example, in 1846, Hugh Edwin Strickland wrote that affinity is "the direct result of those Laws of Organic Life which the Creator has enacted for his own guidance in the act of Creation."[49] According to Ernst Mayr, "Darwin solved this problem by simply stating that affinity is proximity of descent."[50] Darwin also wrote of the affinity of "minute granules or atoms," which he named "gemmules," cast off by cells and "which circulate freely throughout the system, and when supplied with proper nutriment multiply by self-division, subsequently becoming cells like those from which they were derived."[51] Among the many other neologisms suggested by noted nineteenth-century biologists to denote hypothetical protoplasmic constituents were "plastidule" (Ernst Haeckel) and "idio-

47. For a useful summary, see Partington (1964), pp. 608–636.
48. Servos (1990); Fruton (1990), pp. 241–262.
49. Quoted from Mayr (1982), p. 203.
50. Ibid., p. 214.
51. Darwin (1868), vol. 2, p. 374.

plasm" (Carl Nägeli). These words, along with "gemmules," are of considerable historical interest, but have disappeared from the vocabulary of the present-day biologist. On the other hand, many nineteenth-century words, for example "chromatin" (which referred to a cellular element detected by chemical means and microscopic observation), still have a place in contemporary biological language, although the present meaning of such words may not be the same as that originally intended.

It should by now be evident that I do not believe that the introduction of a new word into scientific language, or the use of a word drawn from common speech to denote a newly discovered (or imagined) biochemical constituent or process, has always given, in Huxley's words, "a purer sense to the words of the tribe." All too often, the invention of plausible (and attractive) names for supposedly new biochemical entities appears to have come from the wish of a scientist to assert priority in his discovery. Many of these claims were endorsed in subsequent research, although the original names may have acquired a different meaning, or may have been replaced in later systems of chemical classification; but many other claims were rejected, along with the names given the objects to which they were assigned. Likewise, a new word to denote a relation between biochemical entities may have been accepted, at least for a time, because it appeared to represent a new idea in an attractive way. Some biochemists, notably Willy Kühne and Jacques Monod, were especially adept in inventing such words. In considering such neologisms, however, it should be recalled that a class of biochemical objects or processes may be for a time of interest to investigators in separate areas of scientific inquiry, and that, depending on the nature of the questions under study, different terms may be used for the same objects or processes.

Certainly, short and readily pronounceable words are preferable to polysyllabic jawbreakers, and the use of well-understood abbreviations such as NAD and ATP, or of the three-letter codes for protein amino acids, saves space on the printed page. All of these terms refer to substances of defined chemical structure. One can also appreciate the readiness of biochemical journals to accept as "standard" such terms as ELISA (enzyme-linked immunosorbent assay), PCR (polymerase chain reaction), or PAGE (polyacrylamide gel electrophoresis); these acronyms for widely used methods in present-day biochemical research are unambiguous, although doubts may occasionally arise about the care and skill in the application of these methods in particular investigations. As in the past, however, ambiguities still arise in the languages of the biochemical sciences when new words are introduced to denote biochemical entities which are of unknown (or uncertain) chemical constitution, and which are believed to play a role in particular observed or postulated physiological processes. Such neologisms are frequently based on analogy to words in common usage for familiar objects or phenomena.

The problems inherent in the use of analogy have long been evident to philosophers. For example, in 1745 Pierre Louis Maupertuis wrote: "Analogy frees us from the difficulty of imagining new objects; and of an even greater difficulty, which is to remain in uncertainty. It pleases our mind: but does it please Nature?"[52] I shall not attempt to recount the later history of the philosophical discourse about the uses of analogy in scientific thought.[53] Much of that discourse has dealt with analogies offered in the past in theories about the motion of the molecules of a gas, or the nature of light and sound, and comparison with objects or phenomena of ordinary experience, such as particles of dust or the motion of water waves or the plucking of strings. During the course of the emergence of the biochemical sciences since 1800 there have been numerous other examples of such analogies, some of which have been mentioned previously in this book. A recurrent question has been an updated version of that posed by Maupertuis. Thus, Emil Fischer used the relationship between a key and a lock (both rigid objects) to illustrate his view of the specific interaction of an enzyme with a chemical substance (the "substrate") whose reaction it catalyzes. This analogy has been replaced by various versions (including "jaws") of Daniel Koshland's concept of "induced fit" in order to take account of new experimental evidence for flexibility in the active sites of some enzymes.[54] At the present state of biochemical knowledge, there is general agreement that the catalytic capacity of an enzyme protein depends both on rigidity in most of its polypeptide fabric and flexibility in the region where a substrate is bound and activated by the enzyme. Such changes in the language used to describe a bio-

52. Maupertuis (1756), vol. 2, p. 51. For Maupertuis's biological views, see Roger (1963), pp. 468–487.

53. See Black (1962), Hesse (1966), Ortony (1979), and Kümmel (1989) and references cited therein.

54. Koshland (1958).

chemical process by the use of new words or phrases drawn from ordinary language should be seen, however, in the context of the biochemical thought of our time, and the meaning of these words or phrases is subject to further change in the future. Also, as we have seen in recent times, words derived from other research disciplines, such as communication theory, have acquired meanings rather different from those assigned to them when they were introduced into the vocabulary of molecular genetics.[55]

For historians of the development of the biochemical sciences, the changes in the meaning of surviving or extinct words to denote objects and processes in biological organisms present considerable challenges. Indeed, as in the historical study of other areas of scientific inquiry, the original meaning may be so different from that accepted at a later time as to make the two meanings, to some degree, incommensurable.[56] The difficulties are particularly great in the case of biochemical neologisms which may be considered to belong to both a chemical language and a biological language, for the two languages may have belonged to research specialties which developed by different routes. The separate development of genetics and the chemical study of nucleic acids between 1870 and 1950 provides many examples. Consequently, in examining the transitions in the conceptual framework of an area of biochemical inquiry, it is not sufficient, in my view, to compare developed theories; it is also essential to seek understanding of the meaning of the words used to denote biochemical objects and their relations in the context of these theories and of the experimental knowledge upon which these theories were based.

Is the Scientific Paper a Fraud?

This provocative question was posed in a radio talk by Peter Medawar in 1963. He answered the question in the affirmative because "it misrepresents the processes of thought that accompanied or gave rise to the work that is described in the paper," and he gave the following account of the traditional form of a paper in the biological sciences:

55. Fantini (1988).
56. Kuhn (1983, 1989).

First, there's a section called the 'introduction' in which you merely describe the general field in which your scientific talents are going to be exercised, followed by a section called 'previous work' in which you concede, more or less graciously, that others have dimly groped towards the fundamental truths that you are about to expound. Then a section on 'methods'—that's O.K. Then comes the section called 'results.' The section called 'results' consists of a stream of factual information in which it is considered extremely bad form to discuss the significance of the results you're getting. You have to pretend that your mind is, so to speak, a virgin receptacle, an empty vessel, for information which floods into it from the external world for no reason which you yourself have revealed. You reserve all appraisal of the scientific evidence until the 'discussion' section, and in the discussion you adopt the ludicrous pretense of asking yourself if the information you've collected actually means anything; of asking yourself if any general truths are going to emerge from the contemplation of all the evidence you brandished in the section called 'results.'[57]

This entertaining account, admittedly (in Medawar's words) "rather an exaggeration," is a preface to his echo of Karl Popper's views on the place of induction in the "logic of scientific discovery."[58] Medawar then suggested that

the inductive format of the scientific paper should be discarded. The discussion which in the traditional scientific paper goes last should surely come at the beginning. The scientific facts and scientific acts should follow the discussion, and scientists should not be ashamed to admit, as many of them apparently *are* ashamed to admit, that hypotheses appear in their minds along uncharted by-ways of thought; that they are imaginative and inspirational in character; that they are indeed adventures of the mind. What after all is the good of scientists reproaching others for their neglect of, or indifference to, the scientific style

57. Medawar (1963), p. 377.
58. See Chapter 2, note 5.

of thinking they set such great store by, if their own writings show that they themselves have no clear understanding of it?[59]

Medawar's view that a scientific research article in the biological sciences should reveal the "adventures" of the mind of the author may be seen, in my opinion, not only as a reflection of his philosophical attitude but also as an admirable attempt to "humanize" the literature of science. In recent times others have called attention to the disappearance, from primary research articles, of the words "I" or "we," and to the view that

> scientific journals give practically no clue in our time as to the paths by which discoveries are made or new ideas emerge. To discover these paths we must usually wait till a scientist receives the Nobel Prize or other honor and is willing, from his new eminence, to reveal a little of the unpublished history of his scientific contributions.[60]

In particular, the noted chemist Roald Hoffmann has argued

> . . . for a general humanization of the publication process. Let's relax those strictures, editorial or self-imposed, on portraying in words, in a primary scientific paper, motivation, whether personal or scientific, emotion, historicity, even some of the irrational. So what if it takes a little more space? We *can* keep up with the chemical literature, and tell the mass of hack work from what is truly innovative, without much trouble as it is.[61]

I share wholeheartedly these worthy sentiments, but as a former author of numerous biochemical research reports on what may (or may not) have been "hack work," as a former member of editorial boards of biochemical journals, and now as a skeptical scientist-historian, I must call attention to some aspects of the questions raised by Medawar and Hoffmann.

Perhaps I need not dwell on the fact that the appearance of an individual biological or chemical paper (or set of papers published to-

59. Medawar (1963), p. 378. See Holmes (1987b).
60. Benfey (1959), p. 571.
61. Hoffmann (1988), p. 1602. See also Douglas (1990).

gether) from a particular laboratory is an event within a historical process. As I suggested in the preceding chapter, such a paper (or papers) should be considered in the context of the previous publications of the principal investigator. Some of the separate papers may be deemed to have been "hack work," but when seen in such a larger perspective they may assume greater importance in revealing the scientist's style of research. Nor, perhaps, is it necessary to repeat that the first stage of innovation is imitation, and that in the modern biological and chemical sciences the appreciation of the relative merit of a particular research paper depends largely on the extent of the reader's detailed knowledge of what has been reported previously by other investigators in the same area of inquiry. Moreover, a well-informed student of a biochemical research paper written in the usual formal mode may even discern aesthetic qualities not apparent to other readers.

However laudable the appeals for the "humanization" of the scientific literature may be, I believe that, at the present stage of development of the biochemical sciences, particular emphasis is needed on the responsibilities of an author of a primary research article in periodicals devoted to these sciences. In view of the recent brouhaha about what is euphemistically termed "scientific misconduct," especially in the so-called biomedical sciences, it perhaps necessary to repeat the oft-stated opinion that every person whose name is listed as a coauthor of a scientific paper is responsible for the *validity* of the data reported in that paper.[62] I consider that requirement to apply with special force to the person (or persons, in collaborative efforts of two or more laboratories) who directed the work which led to the publication of that paper. Although a distinction may be made between "mistakes" arising from the limitations of available methods, inadequate control experiments, faulty apparatus, or sloppiness, and the intentional fabrication or "cooking" of experimental data, the principal author (or authors) cannot, in my opinion, be excused from having failed to detect such errors or offenses. Indeed, many instances of "misconduct" have arisen from the wish of a junior associate to provide experimental support for an "adventure" in the mind of the leader of his or her research group. Peter Karlson has cited examples of such behavior in the recent history of the bio-

62. Broad (1981); Broad and Wade (1982); Lodish (1982); Conrad (1982); Chubin (1985).

chemical sciences.[63] It may be argued that if a research article is of sufficient interest to other investigators to warrant a repetition of particular experimental work, the "truth" will come out, but the wasted effort and damage to reputation are no less deplorable.

Another responsibility, often evaded in the past, and no less important at present, is for the author of a primary research article to include honest and accurate statements, in the text and in the list of references, about previously published works on the same or closely related subjects.[64] It would be both unreasonable and impracticable to ask authors to provide an extended chronological account of earlier work on the problem under investigation, although such accounts are indeed necessary in lengthy review articles or in monographs dealing with the historical development of a particular area of inquiry. What is at issue are the author's criteria for the selection of the limited number of citations included in a primary research paper, and the manner in which reference is made to previous work by others on the subject of that paper. One of the most prevalent sources of acrimony among investigators in the biochemical sciences has been the not uncommon wish, on the part of an author, to emphasize the importance of his paper either by ignoring earlier work or, at best, by offering perfunctory (often misleading) references to such work. Moreover, there have been good grounds for the suspicion that some scientists have used knowledge gained during the course of a visit to another laboratory, or in their capacity as "peer reviewers" of a grant application, to repeat someone else's novel experiment and to rush the result into print. Such practices tend to inhibit the exchange of unpublished information even, on occasion, among workers in the same laboratory. These practices may be explained but, in my opinion, not excused by the kind of comparison of the present-day scientific endeavor to capitalist enterprise offered by Richard Kenyon in 1966:

> The young scientist driving for a career that will build a major reputation in his lifetime must get in some successful licks quite early to build a base of prestige useful for getting major grants in his most productive period. This is not too far from the ambitious young financier who must accumulate some capital

63. Karlson (1986).
64. See Hartree (1976) for a case history of inaccurate citation.

> early if ever he is to be able to build a major fortune. . . . Scientific journals must be considered as partially producer-oriented to see them in full light, for the idealistic purpose of serving the scientists who read is not adequate for explaining the total problem. Considering publication only as a service to other scientists is not too unlike considering a businessman or an industrialist being in business only to serve the public.[65]

Although this analogy may indeed apply to many present-day investigators, and "misconduct" in the biomedical sciences may mirror shady business practices to a greater extent than before, the striving for recognition and priority has long been a major feature of the scientific life. In an admirable lecture, Robert Merton recounted some of the relevant history and defined the problem more perceptively:

> The self-contempt often expressed by scientists as they observe with dismay their own concern with having their originality recognized is evidently based upon the widespread though uncritical assumption that behavior is actuated by a single motive, which can then be appraised as 'good' or 'bad,' as noble or ignoble. They assume that the truly dedicated scientist must be concerned only with advancing knowledge. As a result, their deep interest in having their priority recognized by peers is seen as marring their nobility of purpose as men of science (although it might be remembered that 'noble' initially meant the widely-known). This assumption has a germ of psychological truth: any reward—fame, money, position—is morally ambiguous and potentially subversive of culturally esteemed motives. For as rewards are meted out—fame, for example—the motive of seeking the reward can displace the original motive, concern with recognition can displace concern with advancing knowledge. . . . In general, the need to have accomplishment recognized, which for the scientist means that his knowing peers judge his work worth the while, is the result of deep devotion to the advancement of knowledge as an ultimate value. Rather than necessar-

65. Kenyon (1966), p. 5 (Reprinted with permission from *Chemical and Engineering News* (1966) 44: 5; copyright 1966 American Chemical Society.)

> ily being at odds with dedication to science, the concern with recognition is usually a direct expression of it.[66]

To this unexceptionable statement I must add, however, that the question of the moral responsibility to give honest and accurate citations of previous work merits close attention not only in relation to the competition among individual scientists for priority, but also because of the tendency of leading members of some tightly knit research schools in the biochemical sciences to omit or distort references to previous work as a means of delimiting their territory and repelling intruders. As I suggested in the preceding chapter, this tendency is also evident in the literary output of some historians and sociologists who have written about the modern biochemical sciences.

An extensive literature has emerged during the past twenty years on the use of numerical data derived from Eugene Garfield's *Science Citation Index* to determine the quality of the work of a particular scientist or of a particular scientific paper.[67] I shall not attempt to summarize the recent discussion about the value of these efforts, and only compare two statements on this subject. In 1967, the sociologists Stephen Cole and Jonathan Cole wrote that "the invention of the *Science Citation Index* (SCI) a few years ago provides a new tool which yields a reliable and valid measure of the significance of individual scientists' contributions in certain fields of science." In 1989, Stephen Cole asked: "Why are citations a good rough indicator of the quality of work of a sample of scientists and not suitable for comparing the quality of work of individuals? The primary reason is that citations are only a *rough* indicator of quality. In using citations as a measure of quality there is substantial room for error."[68] Apart from the question whether the "substantial room for error" in the use of citation analysis as a measure of "quality" (however that term is defined) is likely to be reduced, one may express skepticism about

66. Merton (1963), pp. 86–87.

67. For an account of the *Science Citation Index*, see Garfield (1974, 1977). Examples of studies in which data from the index were used to gauge the quality and productivity of scientists are Cole and Cole (1967, 1972) and Small (1978). For discussions of the interpretation of such data, see Chubin and Moitra (1975), Gilbert (1977), MacRoberts and MacRoberts (1984, 1989), Cole (1989), and references cited therein.

68. Cole and Cole (1967), p. 379; Cole (1989), p. 11.

the use of the *Science Citation Index* to study other problems of interest to some sociologists. An example of such a problem is the relationship between the "productivity" of individual scientists and the "prestige" of the university departments with which they are affiliated.[69]

In the present social climate of biochemical investigation, especially in the United States, it seems likely that the format of primary research papers will not undergo significant change in the near future, and that the problems related to the clarity and adequacy of exposition and to appropriate citation may be expected to persist. I have little hope that there will be a favorable response to the well-intentioned appeals, in their general writings, of some noted biologists and chemists for less formality and freer expression of individual thought and emotion in primary research reports. My doubts arise in large part from the fact that, to a greater extent than before, research in many lines of chemical and biological investigation is conducted by relatively large and closely directed groups. Not infrequently, the originality of a primary research article has stemmed from the ideas of one or two junior associates, rather than from the group leader whose name appears among the coauthors and who has submitted the paper for publication. Nor, in my opinion, can one expect much improvement in the style of research papers from the recent fashion, fostered by the availability of powerful computers, to use data on the number of papers which included among the coauthors the name of a particular scientist, or on the number of citations of a particular paper in which that name was listed first among its authors. Such efforts may have found favor with managerial bureaucrats in government agencies which dispense funds for scientific research but, like the appeals for the "humanization" of the primary research article, appear to have had little effect in changing the format or style of such articles in the biochemical sciences.

As before, the principal purposes of a research paper in the chemical and biological sciences are twofold: to report what the authors believe to be a contribution to the exploration of a particular area of inquiry, and to direct the attention of their contemporaries to that contribution and to themselves. By the middle of the nineteenth century, with the professionalization of these sciences and the proliferation of specialized periodicals, the literary style largely lost

69. See Allison and Long (1990) and references cited therein.

whatever personal qualities it may have once possessed, and in succeeding decades the admittedly artificial format described by Medawar proved to be an effective means for the transmission of scientific information. In my opinion, the acceleration in the growth of the volume of research publications has not diminished their value either to contemporary or to later scientific investigators (or to historians). For my own part, in reading research reports in many current journals devoted to special areas of organic and physical chemistry, or of biochemistry and molecular biology, I have derived the same exhilaration, as well as occasional exasperation, as from my earlier study of research articles in the now "ancient" literature, because the gap between chemical and biological knowledge still represents a challenging terrain in which new methods and instruments lead to new empirical discovery and theoretical insight.

The vast increase after 1945 in the number of research publications in the biochemical sciences has led some writers to ask whether the scientific literature is worth preserving. For example, in an article reciting the literary deficiencies of scientific research papers, John Maddox concluded that

> the condition of the literature at present is, indeed, so bad that it may well be that most changes would be improvements. Certainly it is ingenuous almost to the point of dishonesty that the scientific community should so persistently badger the librarians for more and more elaborate methods of cataloguing the scientific literature, and for more and more storage space, when it cannot be seen to be doing everything that needs to be done to make the torrent of literature intelligible outside the narrowest of circles.[70]

The prospect that the "torrent" of research papers in the biochemical sciences will be stemmed appears to be dim, for new journals continue to arise for articles not accepted by older periodicals, or for the publication of papers in particular specialties. It is, I believe, safe to say that every article, no matter how meager its scientific content or how inadequate its literary style may be, will now find a place in some periodical if the author is sufficiently determined to publish it. Indeed, the proliferation of new journals in the biochemi-

70. Maddox (1964), p. 178; see Fox (1965).

cal sciences since 1900 has been a consequence not only of the growth of knowledge, increased specialization, and the multiplication of scientific workers, but also of the discontent of investigators with the editorial decisions of established "elite" publications.[71]

Every scientist has learned that in submitting a research article for publication in a respected journal, he or she must expect to receive critical editorial comments. Occasionally a paper may be accepted without a request for revision; more often the reply is in the form of a more or less polite rejection. Few editors of journals in the biochemical sciences have emulated the courtesy reported in an obituary notice for Frank Crowninshield, editor of *Vanity Fair:*

> As an editor Mr. Crowninshield was so kindly towards authors that his rejections sometimes required a second reading for the discovery that they were not acceptances. Once he told Paul Gallico, "My dear Boy: This is supersuperb! A little masterpiece! What color! What life! How beautifully you have phrased it! A veritable gem—why don't you take it around to *Harper's Bazaar?*"[72]

Among the famous rejections in the history of biochemistry was the one received in 1937 by Hans Krebs from *Nature* in regard to his paper about the citric acid cycle:

> The paper was returned to me five days later accompanied by a letter of rejection written in the formal style of those days. This was the first time in my career, after having published more than fifty papers, that I experienced a rejection or semi-rejection. My 'letter' to *Nature* was never published. Instead, two weeks later, I sent the full paper to the journal *Enzymologia* in Holland and it was published within two months.[73]

71. See Webb (1970), Sengupta (1973), Garfield (1979), and Strbanova (1981).

72. *New York Herald Tribune*, 29 December 1947, p. 10. This quotation was included in a letter (2 January 1948) to me from Rudolph Anderson, managing editor of the *Journal of Biological Chemistry*, at the beginning of my service as a member of its editorial board. The letter concluded with the sentence: "I hope that we will accept all the real gems that are submitted to the JBC and return the dross with a kind word."

73. Krebs (1981), pp. 98–99 (quoted by permission of Oxford University Press).

The fact that *Enzymologia* was among the less prestigious biochemical journals of its time (it ceased publication in 1972) did not affect the immediate recognition within the international biochemical community of the importance of Krebs's paper.

Such anecdotal evidence serves only to underline the role played by editors, and their advisers, in the evaluation of the merit of primary research articles. That role has undergone significant changes since the establishment in 1877 of the *Zeitschrift für physiologische Chemie* by Felix Hoppe-Seyler. At first the changes came largely in response to the growing diversity of research tracks, but after World War II they were a consequence of the enormous increase in the number of articles submitted each year to the leading journals in the biochemical sciences. Thus, in 1947 the editorial board of the *Journal of Biological Chemistry* had twelve members, one of whom headed the office which sent out the decision letters that he prepared on advice of his colleagues. Ten years later, to handle the rapidly increasing number and variety of manuscripts, the size of the board had been increased to twenty-six members. In 1990, the editorial board consisted of an "editor," fourteen "associate editors," each of whom informed authors of his decision, and 219 "advisers to the board." Inevitably, this dispersal of editorial responsibility has increased dependence on "peer review" of variable competence, and has decreased the courtesy shown to authors. In former days, a scientific editor, or one of a small group of editors, was expected to know enough of the background of a paper to recognize both its significance and its demerits. If he was relatively free of prejudice, and impressed by the ingenuity of an idea or the plausibility of the argument but unhappy about the experimental evidence offered in their support, a kindly letter suggesting further work was not an uncommon response. On the other hand, if the editor sought to advance, in the pages of his journal, his own views and to put down his scientific adversaries, he was likely to be high-handed in his judgment of papers, especially those by younger investigators. The prime exemplars of this practice in nineteenth-century chemistry were Justus Liebig and Hermann Kolbe.[74] Other editors may have been more generous in their evaluations but, like Arthur Harden of the *Biochemical Journal*, occasionally were somewhat strict in matters of literary style or nomenclature. Moreover, with the growing diversity

74. See Phillips (1966).

of approaches to biochemical problems, some editors often resisted new scientific fashions, and were more inclined to accept papers which made no claim for theoretical significance but were characterized by detailed accounts of experimental methods and results. As a consequence, biochemical journals have contained articles which a contemporary reader may have considered to be "trivial" or "pedestrian." Although such unfavorable opinions may have been justified in many cases, I believe that there ought to be a place in the biochemical literature for reports of work that was well performed, with careful attention to the homogeneity of the substances used and to the precision of the experimental measurements. If we speak of "breakthroughs" and "revolutions" in the development of the biochemical sciences, and cite particular papers of historical importance, we should also recognize that, by and large, this development has been a step-by-step process in which many seemingly pedestrian papers later assumed a greater importance than that anticipated by editors or, in some cases, even by their authors.

It is my impression that, with the dispersal of editorial responsibility, critical judgment has tended to give way to the routine business of collecting the reports of referees and transmitting them verbatim to the author. I shall not venture to judge whether the earlier practice, with its many faults, was more subject to error than the present-day procedure, and only note that, in my opinion, the changes in the responsibility of editors and the increased importance of the peer-review system have not produced an improvement in the format and style of primary research articles in the biochemical sciences.[75]

I have offered these opinions about the languages of the biochemical sciences because the principal themes touched upon in this book—the "scientific method" in the chemical and biological sciences, the reciprocal interactions of chemical and biological investigation, the description and interpretation of the historical development of the biochemical sciences—have in common the problems raised by the ambiguities in the words chosen to denote ideas about natural objects and their relations. With the emergence of new research specialties, new meanings were given to elements of the older scientific language, and new words were invented to denote real or hypotheti-

75. See Zuckerman and Merton (1971); Ziman (1981), p. 31–36; Maddox (1989).

cal objects and phenomena. Occasionally, the same word (for example, "ribonuclease") may have had different meanings to a chemist and to a biologist. The recognition of such ambiguities is, I believe, essential for philosophical discourse about such matters as "reductionism," and for the historical interpretation of the growth of knowledge in the biochemical sciences. In addition, since the invention of new scientific words (or the use of common words in novel ways) is frequently related to the striving for priority or for "discipline building," the changing vocabulary of these sciences is a matter of interest to social historians and to sociologists.

To recognize the problems inherent in the effort to "purify the words of the tribe" does not, in my opinion, require the imposition upon all biochemically-minded scientists of an officially sanctioned vocabulary. Certainly, a rational nomenclature is preferable to one dictated by noted, but idiosyncratic, individual scientists. It would seem, however, that the diversity of specialization within the biochemical sciences has led to differences in language which are not likely to be easily resolved by appeals to reason, especially when the social status (autonomy) of the scientific specialty is at stake. If, at a given time, the terms used to denote particular natural objects and their relations have, in a chemical specialty, meanings which are somewhat different from those in a related biological specialty, perhaps all one may expect is that, as in the past, chemists will be willing to learn the language of biologists (and vice versa), to recognize the differences, and to seek through theory and experiment to narrow the gap between the two specialties. For example, in very recent times, some experimental biologists (geneticists, immunologists, physiologists, pharmacologists) have found it profitable to collaborate with some of the chemists who have specialized in the study of the mechanisms of enzyme-catalyzed reactions. Such developments, in my opinion, are likely to provide the seeds of future transitions in biochemical thought, with resultant further changes in the languages of both the biological and the chemical sciences.

Bibliography
Index of Personal Names
Index of Subjects

Bibliography

Abir-Am, P. (1982a). Modern biochemistry. *British Journal for the History of Science* 15:301–305.

——— (1982b). The discourse of physical power and biological knowledge in the 1930s: A reappraisal of the Rockefeller Foundation's "policy" in molecular biology. *Social Studies of Science* 12:341–382.

——— (1985). Themes, genres, and orders of legitimation in the consolidation of new scientific disciplines: Deconstructing the historiography of molecular biology. *History of Science* 23:73–117.

——— (1987a). Synergy or clash: Disciplinary and marital strategies in the career of the mathematical biologist Dorothy Wrinch. In *Uneasy Careers and Intimate Lives* (P. G. Abir-Am and D. Outram, eds.), pp. 239–280. New Brunswick, N.J.: Rutgers University Press.

——— (1987b). The biotheoretical gathering, trans-disciplinary authority, and the incipient legitimation of molecular biology in the 1930s: New perspective on the historical sociology of science. *History of Science* 25:1–70.

Ackermann, R. J. (1985). *Data, Instruments, and Theory*. Princeton University Press.

Adams, M. B. (1980). Sergei Chetverikov, the Kol'tsov Institute, and the evolutionary synthesis. In *The Evolutionary Synthesis* (E. Mayr and W. B. Provine, eds.), pp. 242–278. Cambridge, Mass.: Harvard University Press.

Adler, J. (1990). Goethe's use of chemical affinity in his *Elective Affinities*. In *Romanticism and the Sciences* (A. Cunningham and N. Jardine, eds.), pp. 263–278. Cambridge University Press.

Akeroyd, F. M. (1988). Research programmes and empirical results. *British Journal for the Philosophy of Science* 39:51–58.

Alborn, T. L. (1989). Negotiating notations: Chemical symbols and British society. *Annals of Science* 46:437–460.

Allen, G. E. (1976). Edmund Beecher Wilson. *Dictionary of Scientific Biography* 14:423–436. New York: Charles Scribner's Sons.

——— (1979). Naturalists and experimentalists: The genotype and the phenotype. *Studies in the History of Biology* 3:179–209.

Allison, P. D., and Long, J. S. (1990). Departmental effects on scientific productivity. *American Sociological Review* 55:469–478.

Altman, S. (1990). Enzymatic cleavage of RNA by RNA. *Angewandte Chemie (International Edition)* 29:749–758.

Amsterdamska, O. (1987). Medical and biological constraints: Early research on variation in bacteriology. *Social Studies of Science* 17:657–687.

Anderson, W. C. (1984). *Between the Library and the Laboratory: The Language of Chemistry in Eighteenth-Century France*. Baltimore: The Johns Hopkins University Press.

Andrews, F. M. (ed.). (1979). *Scientific Productivity: The Effectiveness of Research Groups in Six Countries*. Cambridge University Press.

Anfinsen, C. B., and Haber, E. (1961). Studies on the reduction and reformation of protein disulfide bonds. *Journal of Biological Chemistry* 236:1361–1363.

Anthony-Cahill, S. J., Griffith, M. C., Noren, C. J., Suich, D. J., and Schultz, P. G. (1989). Site-specific mutagenesis with unnatural amino acids. *Trends in Biochemical Sciences* 14:400–403.

Aris, R., Davis, H. T., and Steuwer, R. H. (eds.). (1983). *Springs of Scientific Creativity*. Minneapolis: University of Minnesota Press.

Arthus, M. (1921). *De l'Anaphylaxie à l'Immunité*. Paris: Masson.

Astbury, W. T. (1952). Adventures in molecular biology. *Harvey Lectures* 46:3–44.

Atkinson, D. E. (1965). Biological feedback control at the molecular level. *Science* 150:851–857.

Auerbach, C. (1967). The chemical production of mutations. *Science* 158:1141–1147.

Avery, O. T., MacLeod, C. M., and McCarty, M. (1944). Studies on the chemical nature of the substance inducing transformation of pneumococcal types. *Journal of Experimental Medicine* 79:137–158.

Ayala, F. J. (1968). Biology as an autonomous science. *American Scientist* 56:207–221.

——— (1974). Introduction. In *Studies in the Philosophy of Biology* (F. J. Ayala and T. Dobzhansky, eds.), pp. vii–xvi. Berkeley: University of California Press.

Ayala, F. J., et al. (1989). *On Being a Scientist*. Washington, D.C.: National Academy Press.

Ayer, A. J. (1952). *Language, Truth and Logic*. New York: Dover.

——— (1963). Professor Popper's work in progress. *New Statesman*, 1 February, pp. 155–156.

Bachelard, G. (1938). *La Formation de l'Esprit Scientifique*. Paris: Vrin.

——— (1972). *Le Matérialisme Rationnel*. Paris: Presses Universitaires de France.

——— (1973). *Le Pluralisme Cohérent de la Chimie Moderne*. Paris: Vrin.

Bacon, F. (1884). *The Works of Francis Bacon* (B. Montagu, ed.). Second ed. New York: Worthington.

Baird, D., and Faust, T. (1990). Scientific instruments, scientific progress, and the cyclotron. *British Journal for the Philosophy of Science* 41:147–175.

Baker, J. R. (1948–1955). The cell-theory: A restatement, history, and critique. *Quarterly Journal of Microscopical Science* 89:103–125, 90:87–108, 93:157–190, 94:407–440, 96:449–481.

Baldwin, E. (1952). *Dynamic Aspects of Biochemistry*. Second ed. Cambridge University Press.

Baltzer, F. (1962). *Theodor Boveri*. Stuttgart: Wissenschaftliche Verlagsgesellschaft. English translation (1967) by D. Rudnick. Berkeley: University of California Press.

Balzac, H. de (1925). *Oeuvres Complètes de Honoré de Balzac. La Comédie Humaine. Études Philosophiques I* (M. Bouteron and H. Longonon, eds.). Paris: Louis Conard.

Barcroft, J. (1928). *The Respiratory Function of the Blood, Part II, Haemoglobin*. Cambridge University Press.

Barford, D., and Johnson, L. N. (1989). The allosteric transition of glycogen phosphorylase. *Nature* 340:609–616.

Baron, J., and Polanyi, M. (1913). Ueber die Anwendigkeit des zweiten Hauptsatzes der Thermodynamik auf Vorgänge im tierischen Organismus. *Biochemische Zeitschrift* 53:1–20.

Bateson, W. (1909). *Mendel's Principles of Heredity*. Cambridge University Press.

——— (1913). *Problems of Genetics*. New Haven: Yale University Press.

——— (1916). The mechanism of Mendelian heredity. *Science* 44:536–543.

Bawden, F. C., Pirie, N. W., Bernal, J. D., and Fankuchen, I. (1936). Liquid crystalline substances from virus infected plants. *Nature* 138:1051–1052.

Baxter, A. L. (1974). *Edmund Beecher Wilson and the Problem of Development*. New Haven: Yale University Press.

Bayliss, W. M. (1924). *Principles of General Physiology*. Fourth ed. London: Longmans Green.

Beadle, G. W. (1963). *Genetics and Modern Biology*. Philadelphia: American Philosophical Society.

Beatty, J., and Finsen, S. (1989). Rethinking the propensity interpretation: A peek inside Pandora's box. In *What the Philosophy of Biology Is* (M. Ruse, ed.), pp. 17–30. Dordrecht: Kluwer.

Becher, T. (1989). *Academic Tribes and Territories: Intellectual Enquiry and the Culture of Disciplines*. Milton-Keynes: Open University Press.

Bechtel, W. (1986). *Integrating Scientific Disciplines*. Dordrecht: Nijhoff.

Beckner, M. (1969). Function and teleology. *Journal of the History of Biology* 2:151–164.

——— (1971). Teleology. In *Man and Nature* (R. Munson, ed.), pp. 92–101. New York: Dell.

Bédarida, F. (1987). The modern historian's dilemma: Conflicting pressures from science and society. *Economic History Review* [2] 40:335–348.

Beijerinck, M. W. (1899). Ueber ein Contagium vivum fluidum als Ursache der Fleckenkrankheit der Tabaksblätter. *Centralblatt für Bakteriologie, Parasitenkunde und Infektionskrankheiten* II. Abteilung 5:27–33.

Belt, H. van den, and Gremmen, B. (1990). Specificity in the era of Koch and Ehrlich: A generalized interpretation of Ludwik Fleck's "serological" thought style. *Studies in the History and Philosophy of Science* 21:463–479.

Bender, M., and Kezdy, F. J. (1965). Mechanism of the action of proteolytic enzymes. *Annual Review of Biochemistry* 34:49–76.

Benfey, O. T. (1959). A. W. Williamson and the impersonal passive. *Journal of Chemical Education* 36:571.

Benner, S., and Allemann, R. K. (1989). The return of pancreatic ribonucleases. *Trends in Biochemical Sciences* 14:396–397.

Benton, E. (1974). Vitalism in nineteenth-century scientific thought: A typology and reassessment. *Studies in the History and Philosophy of Science* 5:17–48.

Bergmann, M. (1935). Complex salts of amino acids and peptides. II. Determination of *l*-proline with the aid of rhodanilic acid and the structure of gelatin. *Journal of Biological Chemistry* 110:471–479.

Bergmann, M., and Niemann, C. (1937). On the structure of proteins: cattle hemoglobin, egg albumin, cattle fibrin, and gelatin. *Journal of Biological Chemistry* 118:301–314.

Bergmann, M., and Zervas, L. (1932). Ueber ein allgemeines Verfahren der Peptidsynthese. *Berichte der deutschen chemischen Gesellschaft* 65:1192–1201.

Berlinski, D. (1976). On Systems Analysis. Cambridge, Mass.: The MIT Press.

Bernal, J. D. (1930). The place of X-ray crystallography in the development of modern science. *Radiology* 15:1–12.

——— (1963). William Thomas Astbury. *Biographical Memoirs of Fellows of the Royal Society* 9:1–35.

——— (1968). The material basis of life. *Labour Monthly*, July, pp. 323–326.

Bernard, C. (1857). Sur le mécanisme physiologique de la formation du sucre dans le foie. *Comptes Rendus de l'Académie des Sciences* 44:578–586.

——— (1865). *Introduction a l'Étude de la Médecine Expérimentale*. Paris: Ballière.

——— (1877). *Leçons sur le Diabète et la Glycogenèse Animale*. Paris: Ballière.

——— (1879). *Leçons sur les Phénomènes de la Vie commun aux Animaux et aux Végétaux*. Paris: Ballière.

——— (1947). *Principes de Médecine Expérimentale*. Paris: Presses Universitaires de France.

Berzelius, J. J. (1833–1841). *Lehrbuch der Chemie*. Third ed.; German translation by F. Wöhler. Dresden: Arnold.

Beveridge, W. I. B. (1951). *The Art of Scientific Investigation*. New York: W. W. Norton.

Black, M. (1954). *Problems of Analysis*. Ithaca, N.Y.: Cornell University Press.

——— (1962). *Models and Metaphors*. Ithaca, N.Y.; Cornell University Press.

Bloch, M. (1954). *The Historian's Craft*. Manchester, Eng.: Manchester University Press.

——— (1961). *Feudal Society*. University of Chicago Press.

Blow, D. M., Fersht, A. R., and Winter, G. (eds.) (1986). *Design, Construction, and Properties of Novel Protein Molecules*. London: Royal Society.

Bodansky, M. (1938). *Introduction to Physiological Chemistry*. Fourth ed. New York: Wiley.

Bollum, F. J. (1981). How to find your niche in twenty-first century molecular biology. *Trends in Biochemical Sciences* 6(11): iii–v.

Bouchard, G. (1938). *Guyton-Morveau, Chimiste et Conventionnel*. Paris: Perrin.

Boyle, R. (1661). *The Sceptical Chymist*. London: Cadwell and Crooke.

Bradbury, S. (1967). *The Evolution of the Microscope*. Oxford: Pergamon.

Bragg, W. L., Kendrew, J. C., and Perutz, M. F. (1950). Polypeptide chain configurations in crystalline proteins. *Proceedings of the Royal Society* A203:321–357.

Brandén, C. I., and Jones, T. A. (1990). Between objectivity and subjectivity. *Nature* 343:687–689.

Bredt, D. S., and Snyder, S. H. (1990). Isolation of nitric oxide synthetase, a calmodulin-requiring enzyme. *Proceedings of the National Academy of Sciences* 87:682–685.

Bremer, K., Bremer, B., Karis, P. O., and Kallersjö, M. (1990). Time for a change in taxonomy. *Nature* 343:202.

Brenner, S., Jacob, F., and Meselson, M. (1961). An unstable intermediate carrying information from genes to ribosomes for protein synthesis. *Nature* 190:576–581.

Brillouin, L. (1962). *Science and Information Theory*. New York: Academic Press.

Brink, R. A. (ed.) (1967). *Heritage from Mendel*. Madison, Wis.: University of Wisconsin Press.

Broad, W. J. (1981). Fraud and the structure of science. *Science* 212: 137–141.

Broad, W. J., and Wade, N. (1982). *Betrayers of the Truth*. New York: Simon and Schuster.

Brooke, J. H. (1971). Organic synthesis and the unification of chemistry. *British Journal of the History of Science* 5:363–392.

Brooks, C. McC., and Cranefield, P. F. (eds.) (1959). *The Historical Development of Physiological Thought*. New York: Hafner.

Brücke, E. (1861). Die Elementarorganismen. *Sitzungsberichte der kaiserlichen Akademie der Wissenschaften in Wien, Mathematische-Naturwissenschaftliche Classe* 44(2): 381–406.

Bruice, T. C. (1976). Some pertinent aspects of mechanism as determined with small molecules. *Annual Review of Biochemistry* 45:331–375.

Buchdahl, G. (1973). Leading principles and induction: The methodology of Matthias Schleiden. In *Foundations of Scientific Method: The Nineteenth Century* (R. N. Giere and R. S. Westfall, eds.), pp. 23–52. Bloomington: Indiana University Press.

Bulloch, W. (1938). *The History of Bacteriology*. London: Oxford University Press.

Bunge, M. (1961). The weight of simplicity in the construction and assaying of scientific theories. *Philosophy of Science* 28:120–149.

Burchardt, L. (1978). Die Ausbildung des Chemikers im Kaiserreich. *Tradition* 12:31–53.

——— (1980). Professionalisierung oder Berufskonstruktion? Das Beispiel des Chemikers im wilhelminischen Deutschland. *Geschichte und Gesellschaft* 6:326–348.

Burckhardt, J. J. (1988). *Die Symmetrie der Kristalle*. Basel: Birkhäuser.

Burian, R. (1985). On conceptual change in biology: The case of the gene. In *Evolution at the Crossroads: The New Biology and the New Philosophy of Science* (D. J. Depew and B. H. Weber, eds.), pp. 21–42. Cambridge, Mass.: The MIT Press.

Burke, J. G. (1966). *Origins of the Science of Crystals*. Berkeley: University of California Press.

Butler, A. R. (1990). NO—Its role in the control of blood pressure. *Chemistry in Britain* 26: 419–421.

Bykov, G. V. (1962). The origin of the theory of chemical elements. *Journal of Chemical Education* 39:220–224.

Bynum, W. F. (1970). Chemical structure and pharmacological action: A chapter in the history of molecular pharmacology. *Bulletin of the History of Medicine* 44:518–538.

Cahn, R. S., and Dermer, O. C. (1979). *Introduction to Chemical Nomenclature*. Fifth ed. London: Butterworths.

Cairns, J., Stent, G. S., and Watson, J. D. (eds.) (1966). *Phage and the Origins of Molecular Biology*. Cold Spring Harbor, N.Y.: Laboratory of Quantitative Biology.

Callender, L. A. (1988). Gregor Mendel: An opponent of descent with modification. *History of Science* 26:41–75.

Canguilhem, G. (1968). *Études d'Histoire et de Philosophie des Sciences*. Paris: Vrin.

——— (1977). *Idéologie et Rationalité dans l'Histoire des Sciences de la Vie*. Paris: Vrin.

Carlson, E. A. (1966). *The Gene: A Critical History*. Philadelphia: Saunders.

Caron, J. A. (1988). "Biology" in the life sciences: A historiographical contribution. *History of Science* 26:223–268.

Carpenter, F. H. (1960). The free energy change in hydrolytic reactions. The non-ionized compound convention. *Journal of the American Chemical Society* 82:1111–1122.

Carter, C. W. (ed.) (1990). Protein and nucleic acid crystallization. *Methods* 1:1–104.

Caruthers, M. H. (1985). Gene synthesis machines: DNA chemistry and its uses. *Science* 230:281–285.

Cassebaum, H. (1971). Die Stellung der Arbeiten von J. B. Dumas (1800–1884) und A. Strecker (1822–1871) in der Entwicklung des Periodensystems. *NTM* 8:46–57.

Cassebaum, H., and Kauffman, G. B. (1971). The periodic system of the chemical elements. The search for its discoverer. *Isis* 62:314–327.

Chantrenne, H. (1966). For the 25th anniversary of ~P. In *Current Aspects of Biochemical Energetics* (N. O. Kaplan and E. G. Kennedy, eds.), pp. 33–37. New York: Academic Press.

Chargaff, E. (1950). Chemical specificity of nucleic acids and mechanisms of their enzymatic degradation. *Experientia* 6:201–209.

——— (1963). *Essays on Nucleic Acids*. Amsterdam: Elsevier.

——— (1968). A quick climb up Mount Olympus. *Science* 159:1448–1449.

——— (1974). Building the tower of babble. *Nature* 248:776–779.

——— (1978). *Heraclitean Fire*. New York: Rockefeller University Press.

——— (1986). *Serious Questions*. Boston: Birkhäuser.

——— (1989). In retrospect. *Biochimica et Biophysica Acta* 1000:15–16.

Chibnall, A. C. (1942). Amino-acid analysis and the structure of proteins. *Proceedings of the Royal Society* B131:136–160.

——— (1966). The road to Cambridge. *Annual Review of Biochemistry* 35:1–22.

——— (1987). *Early Days in Biochemistry*. London: Biochemical Society.

Chubin, D. E. (1976). The conceptualization of scientific disciplines. *Sociological Quarterly* 17:448–476.

——— (1985). Misconduct in research: An issue in science policy and practice. *Minerva* 23:175–202.

Chubin, D. E., and Moitra, S. D. (1975). Content analysis of references: Adjunct or alternative to citation counting? *Social Studies of Science* 5:423–441.

Churchill, F. B. (1969). From machine-theory to entelechy: Two studies in developmental physiology. *Journal of the History of Biology* 2:165–185.

——— (1974). Wilhelm Johannsen and the genotype concept. *Journal of the History of Biology* 7:5–30.

Churchill, F. B., et al. (1989). Toward the history of protozoology. *Journal of the History of Biology* 22:185–356.

Claesson, S., and Pedersen, K. O. (1972). The Svedberg. *Biographical Memoirs of Fellows of the Royal Society* 18:595–627

Clark, B. R. (ed.) (1987). *The Academic Profession.* Berkeley: University of California Press.

Clark, G., and Kasten, F. H. (1983). *History of Staining.* Baltimore: Williams and Wilkins.

Clementi, E., Coronglu, G., Sarma, M. H., and Sarma, R. H. (1985). *Structure and Motion: Membranes, Nucleic Acids, and Proteins.* Guilderland, N.Y.: Adenine Press.

Clericuzio, A. (1990). A redefinition of Boyle's chemistry and corpuscular philosophy. *Annals of Science* 47:561–589.

Cohen, G. N. (1986). Four decades of Franco-American collaboration in biochemistry and molecular biology. *Perspectives in Biology and Medicine* 29:S141-S148.

Cohen, I. B. (1956). *Franklin and Newton.* Philadelphia: American Philosophical Society.

——— (1985). *Revolution in Science.* Cambridge, Mass.: Harvard University Press.

——— (1990). *Benjamin Franklin's Science.* Cambridge. Mass.: Harvard University Press.

Cohen, L. J. (1989). *An Introduction to the Philosophy of Induction and Probability.* Oxford University Press.

Cohen, P., and Cohen, P. T. W. (1989). Protein phosphatases come of age. *Journal of Biological Chemistry* 264:21435–21438.

Cohen, R. S., and Toulmin, S. (eds.) (1986). *Cognition and Fact: Materials on Ludwik Fleck.* Dordrecht: Reidel.

Cohen, S. S. (1975). The origins of molecular biology. *Science* 187:827–830.

——— (1984). The biochemical origins of molecular biology. *Trends in Biochemical Sciences* 9:334–336.

Cole, J. R., and Cole, S. (1972). The Ortega hypothesis. *Science* 178:368–375.

Cole, S. (1989). Citations and the evaluation of individual scientists. *Trends in Biochemical Sciences* 14:9,11,13.

Cole, S., and Cole, J. R. (1967). Scientific output and recognition: A study in the operation of the reward system in science. *American Sociological Review* 32:377–390.

Cole, S., Cole, J. R., and Simon, G. A. (1981). Chance and consensus in peer review. *Science* 214:881–886.

Coleman, W. (1965). Cell, nucleus, and inheritance: An historical study. *Proceedings of the American Philosophical Society* 109:124–158.

Collard, P. (1976). *The Development of Microbiology*. Cambridge University Press.
Colvin, J. R., Smith, D. B., and Cook, W. H. (1954). The microheterogeneity of proteins. *Chemical Reviews* 54:687–711.
Conant, J. B. (1950). *Robert Boyle's Experiments in Pneumatics*. Cambridge, Mass.: Harvard University Press.
Conklin, E. G. (1905). Organ-forming substances in the eggs of Ascidians. *Biological Bulletin* 8:205–230.
Conrad, G. W. (1982). Authorship and responsibility in scientific publications and manuscript reviews. *Trends in Biochemical Sciences* 7:167–168.
Cooper, W. (1953). *The Struggles of Albert Woods*. Garden City, N.Y.: Doubleday.
Costabel, P. (1978). Siméon Denis Poisson. *Dictionary of Scientific Biography* 15:480–490. New York: Charles Scribner's Sons.
Coulson, C. A. (1961). *Valence*. Second ed. Oxford University Press.
Crabtree, B., and Taylor, D. J. (1979). Thermodynamics and metabolism. In *Biochemical Thermodynamics* (M. N. Jones, ed.), pp. 333–378. Amsterdam: Elsevier.
Cracraft, J. (1989). Species as entities of biological theory. In *What the Philosophy of Biology Is* (M. Ruse, ed.), pp. 31–52. Dordrecht: Kluwer.
Cranefield, P. F. (1957). The organic physics of 1847 and the biophysics of today. *Journal of the History of Medicine* 12:407–423.
Crick, F. (1958). On protein synthesis. *Symposia of the Society of Experimental Biology* 12:138–163.
——— (1966). *Of Molecules and Men*. Seattle: University of Washington Press.
——— (1974). The double helix: A personal view. *Nature* 248:766–769.
——— (1988). *What Mad Pursuit*. New York: Basic Books.
Crombie, A. C. (1963). *Scientific Change*. London: Heinemann.
Crosland, M. P. (1962). *Historical Studies in the Language of Chemistry*. London: Heinemann.
Crowther, J. G. (1974). *The Cavendish Laboratory 1874–1974*. New York: Science History Publications.
Culp, S., and Kitcher, P. (1989). Theory structure and theory change in contemporary molecular biology. *British Journal for the Philosophy of Science* 40:459–483.
Dagognet, F. (1969). *Tableaux et Langages de la Chimie*. Paris: Seuil.
Dale, H. H. (1955). Edward Mellanby. *Biographical Memoirs of Fellows of the Royal Society* 1:193–222.
Darwin, C. (1868). *The Variation of Animals and Plants under Domestication*. New York: Orange Judd.
Davies, M. (1990). W. T. Astbury, Rosie Franklin, and DNA: A memoir. *Annals of Science* 47:607–618
Davis, B. D. (1980). The double helix. *American Scientist* 70:76–77.

Dean, A. C. R., and Hinshelwood, C. D. (1966). *Growth, Function, and Regulation in Bacterial Cells*. Oxford University Press.

Dear, P. (ed.) (1991). *The Literary Structure of Scientific Thought*. Philadelphia: University of Pennsylvania Press.

Debru, C. (1983). *L'Esprit des Protéines*. Paris: Hermann.

De Duve, C. (1988). Prebiotic syntheses and the mechanism of early chemical evolution. In *The Roots of Modern Biochemistry* (H. Kleinkauf, H. von Döhren, and L. Jaenicke, eds.), pp. 881–894. Berlin: de Gruyter.

De Kruif, P. (1962). *The Sweeping Wind*. New York: Harcourt, Brace & World.

Delbrück, M. (1941). A theory of autocatalytic synthesis of polypeptides and its application to the problem of chromosome replication. *Cold Spring Harbor Symposia on Quantitative Biology* 9:122–126.

——— (1949). A physicist looks at biology. *Transactions of the Connecticut Academy of Sciences* 38:175–191.

——— (1971). Aristotle-totle-totle. In *Of Microbes and Life* (J. Monod and E. Borek, eds.), pp. 50–55. New York: Columbia University Press.

——— (1986). *Mind from Matter?* Palo Alto: Blackwell.

Denbigh, K. G. (1989a). Note on entropy, disorder, and disorganization. *British Journal for the Philosophy of Science* 40:323–332.

——— (1989b). The many faces of irreversibility. *British Journal for the Philosophy of Science* 40:501–518.

Denbigh, K. G., and Denbigh, J. S. (1985). *Entropy in Relation to Incomplete Knowledge*. Cambridge University Press.

Dennstedt, M. (1899). Die Entwicklung der organischen Elementaranalyse. *Sammlung chemischer und chemisch-technischer Vorträge* 4:1–114.

Dickerson, R. E., and Geis, I. (1983). *Hemoglobin: Structure, Function, Evolution and Pathology*. Menlo Park, Calif.: Benjamin & Cummings.

Dienert, F. V. (1900). *Sur la Fermentation du Galactose et sur l'Accoutumance des Levures à ce Sucre*. Sceaux: Charaire.

Dixon, M., and Webb, E. C. (1958). *Enzymes*. New York: Academic Press.

Döhren, H. von and Kleinkauf, H. (1988). Research on nonribosomal systems: Biosynthesis of peptide antibiotics. In *Roots of Modern Biochemistry* (H. Kleinkauf, H. von Döhren, and L. Jaenicke, eds.), pp. 355–367. Berlin: de Gruyter.

Dolby, R. G. A. (1984). Thermochemistry versus thermodynamics: The nineteenth century. *History of Science* 22:375–400.

Dolman, C. E. (1973). Robert Koch. *Dictionary of Scientific Biography* 7:420–435. New York: Charles Scribner's Sons.

Donohue, J. (1976). Honest Jim? *Quarterly Review of Biology* 51:285–289.

Douglas, C. (1990). A revolt in style. *Chemistry in Britain* 26:773–775.

Drabkin, D. L. (1975). *Fundamental Structure: Nature's Architecture*. Philadelphia: University of Pennsylvania Press.

Driesch, H. (1894). *Analytische Theorie der organischen Entwicklung*. Leipzig: Engelmann.

Dubos, R. J. (1945). *The Bacterial Cell*. Cambridge, Mass.: Harvard University Press.

——— (1976). *The Professor, the Institute, and DNA*. New York: Rockefeller University Press.

Ducasse, C. J. (1960). Francis Bacon's philosophy of science. In *Theories of Scientific Method* (E. H. Madden et al., eds.), pp. 50–74. Seattle: University of Washington Press.

Dumas, J. B. (1837). *Leçons sur la Philosophie Chimique*. Paris: Ebrard.

Dunn, L. C. (1965). *A Short History of Genetics*. New York: McGraw-Hill.

——— (1969). Genetics in historical perspective. In *Genetic Organization* (E. W. Caspari and A. W. Ravin, eds.), pp. 1–90. New York: Academic Press.

Du Vigneaud, V. (1952). *A Trail of Research*. Ithaca, N.Y.: Cornell University Press.

Edsall, J. T. (1972). Blood and hemoglobin: The evolution of knowledge of function and adaptation in a biochemical system. *Journal of the History of Biology* 5:205–257.

——— (1980). Horace Judson and the molecular biologists. *Journal of the History of Biology* 13:141–158.

——— (1982). Progress in our understanding of biology. In *Progress and its Discontents* (G. A. Almond, M. Chodorow, and R. H. Pearce, eds.), pp. 135–160. Berkeley: University of California Press.

——— (1983). J. D. Bernal, R. H. Fowler, and the nature of water. *Trends in Biochemical Sciences* 8:30–31.

——— (1985). Jeffries Wyman and myself: A story of two interacting lives. In *Selected Topics in the History of Biochemistry* (G. Semenza, ed.), pp. 99–195. Amsterdam: Elsevier.

——— (1986). Understanding blood and hemoglobin: An example of international relations in science. *Perspectives in Biology and Medicine* 29:S107-S123.

——— (1990). Jeffries Wyman: Scientist, philosopher, and adventurer. *Biophysical Chemistry* 37:7–14.

Edsall, J. T., and Wyman, J. (1958). *Biophysical Chemistry*. New York: Academic Press.

Ehrlich, P. (1910). *Studies in Immunity*. Second ed. New York: Wiley.

Eigen, M., and Schuster, P. (1979). *The Hypercycle*. Heidelberg: Springer.

Eklund, J. (1975). *The Incompleat Chymist: Being an Essay on the Eighteenth-Century Chemist in his Laboratory, with a Dictionary of the Obsolete Chemical Terms of the Period*. Washington, D.C.: Smithsonian Institution.

Elias, N., Martins, H., and Whitley, R. (1982). *Scientific Establishments and Hierarchies*. Dordrecht: Reidel.

Elliott, T. R. (1933). Walter Morley Fletcher. *Obituary Notices of Fellows of the Royal Society* 1:153–163.

Ellis, E., and Delbrück, M. (1939). The growth of bacteriophage. *Journal of General Physiology* 22:365–384.
Elton, G. R. (1967). *The Practice of History*. New York: Crowell.
Ephrussi, B. (1953). *Nucleo-cytoplasmic Relations in Micro-organisms*. Oxford University Press.
Erlenmeyer, E., and Schöffer, A. (1859). Ein experimentellkritischer Beitrag zur Kenntnis der Eiweisskorper. *Zeitschrift für Chemie* 2:315–343.
Eschenmoser, A., and Wintner, C. E. (1977). Natural product synthesis and vitamin B_{12}. *Science* 196:1410–1420.
Euler, H. von (1952). In van't Hoff's laboratory in Berlin 1899 and 1900. *Chemisch Weekblad* 48:644–645.
Ewald, P. P. (ed.) (1962). *Fifty Years of X-Ray Crystallography*. Utrecht: International Union of Crystallography.
Fantini, B. (1984). Chemical and biological classification of proteins. *History and Philosophy of the Life Sciences* 5:3–32.
——— (1988). Utilisation par la genetique moléculaire du vocabulaire de la théorie de l'information. In *Transfert de Vocabulaire dans les Sciences* (M. Groult, P. Louis, and J. Roger, eds.), pp. 159–170. Paris: Centre National de la Recherche Scientifique.
Farrington, B. (1953). On misunderstanding the philosophy of Francis Bacon. In *Science, Medicine, and History* (E. A. Underwood, ed.), vol. 1, pp. 439–450. Oxford University Press.
Fasman, G. (ed.) (1989). *Prediction of Protein Structure and the Principles of Protein Conformation*. New York: Plenum.
Fersht, A. R. (1988). Relationships between apparent binding energies measured in site-directed mutagenesis experiments and energetics of binding and catalysis. *Biochemistry* 27:1577–1587.
Feyerabend, P. (1975). *Against Method*. London: NLB.
——— (1987). *Farewell to Reason*. London: Verso.
Fieser, L. F. (1964). *The Scientific Method*. New York: Reinhold.
Fink, C. (1989). *Marc Bloch: A Life in History*. Cambridge University Press.
Fischer, D. H. (1970). *Historians' Fallacies*. New York: Harper & Row.
Fischer, E. (1906). *Untersuchungen über Aminosäuren, Polypeptide und Proteine I (1899–1906)*. Berlin: Springer.
——— (1907). *Untersuchungen in der Puringruppe*. Berlin: Springer.
——— (1916). Isomerie der Polypeptide. *Sitzungsberichte der Preussischen Akademie der Wissenschaften zu Berlin*, pp. 990–1008.
——— (1923). *Untersuchungen über Aminosäuren, Polypeptide und Proteine II (1907–1919)* (M. Bergmann, ed.). Berlin: Springer.
——— (1924). *Untersuchungen aus verschiedenen Gebieten* (M. Bergmann, ed.). Berlin: Springer.
——— (1987). *Aus meinem Leben* (M. Bergmann, ed.). Reprint of 1922 edition with foreword by B. Witkop. Berlin: Springer.
——— (1955). Justus Liebig und Wilhelm Ostwald. *Naturwissenschaftliche Rundschau* 8:49–53.

Fischer, E. P. (1985). *Licht und Leben. Ein Bericht über Max Delbrück, den Wegbereiter der Molekularbiologie.* Constance: Universitätsverlag.
——— (1987). Biologie und Philosophie oder: die andere Gesetze der Physik. *Physikalische Blätter* 43:337–339.
Fischer, E. P., and Lipson, C. (1988). *Thinking about Science: Max Delbrück and the Origins of Molecular Biology.* New York: Norton.
Fisher, N. W. (1973a). The nature of the chemical atom. *History of Science* 11:53–61.
——— (1973b). Organic classification before Kekulé. *Ambix* 20:106–131, 209–233.
——— (1974). Kekulé and organic classification. *Ambix* 21:29–52.
Fleck, L. (1935). *Entstehung und Entwicklung einer wissenschaftlichen Tatsache.* Basel: Benno Schwalbe.
——— (1979). *Genesis and Development of a Scientific Fact* (T. J. Trenn and R. K. Merton, eds.; translation of Fleck [1935]). University of Chicago Press.
Fletcher, M. (1957). *The Bright Countenance: A Personal Biography of Walter Fletcher.* London: Hodder & Stoughton.
Fletcher, W. M., and Hopkins, F. G. (1907). Lactic acid and amphibian muscle. *Journal of Physiology* 35:247–309.
Flood, W. E. (1963). *The Origins of Chemical Names.* London: Oldbourne.
Florkin, M. (1972). *A History of Biochemistry. I. Proto-biochemistry. II. From Proto-biochemistry to Biochemistry.* Amsterdam: Elsevier.
——— (1979). *A History of Biochemistry. V. The Unravelling of Bio-synthetic Pathways.* Amsterdam: Elsevier.
Forbes, R. J. (1948). *Short History of the Art of Distillation.* Leiden: Brill.
Forman, P. (1971). Weimar culture, causality and quantum theory, 1918–1927. *Historical Studies in the Physical Sciences* 3:1–115.
Fourcroy, A. F. (1801–1802). *Système des Connaissances Chimiques.* Third ed. Paris: Baudouin.
Fox, S. W., and Dose, K. (1972). *Molecular Evolution and the Origin of Life.* San Francisco: Freeman.
Fox, T. (1965). *Crisis in Communication.* London: Athlone Press.
Franklin, A. (1986). *The Neglect of Experiment.* Cambridge University Press.
Freudenthal, H. (1976). Norbert Wiener. *Dictionary of Scientific Biography* 14:344–347. New York: Charles Scribner's Sons.
Freund, I. (1904). *The Study of Chemical Composition.* Cambridge University Press.
Fruton, J. S. (1951). The place of biochemistry in the university. *Yale Journal of Biology and Medicine* 23:305–310.
——— (1963). Chemical aspects of protein synthesis. In *The Proteins* (H. Neurath, ed.), vol. 1, pp. 189–310. New York: Academic Press.
——— (1972). *Molecules and Life: Historical Essays on the Interplay of Chemistry and Biology.* New York: Wiley.

——— (1975). Review of R. Olby, *The Path to the Double Helix*. *Journal of the History of Medicine* 30:284–285.

——— (1976). The emergence of biochemistry. *Science* 192:327–334.

——— (1977). Some aspects of biochemical catalysis. *Proceedings of the American Philosophical Society* 121:309–315.

——— (1979a). Claude Bernard the scientist. In *Claude Bernard and the Internal Environment* (E. D. Robin, ed.), pp. 35–41. New York: Dekker.

——— (1979b). Early theories of protein structure. *Annals of the New York Academy of Sciences* 125:1–15.

——— (1982a). The education of a biochemist. In *Of Oxygen, Fuels and Living Matter* (G. Semenza, ed.), pp. 315–354. Wiley: Chichester.

——— (1982b). Proteinase-catalyzed synthesis of peptide bonds. *Advances in Enzymology* 53:239–306.

——— (1982c). *A Bio-bibliography for the History of the Biochemical Sciences since 1800*. Philadelphia: American Philosophical Society.

——— (1983). Review of R. E. Kohler, *From Medical Chemistry to Biochemistry*. *American Scientist* 71:93–94.

——— (1985). Contrasts in scientific style. Emil Fischer and Franz Hofmeister: Their research groups and their theory of protein structure. *Proceedings of the American Philosophical Society* 129:313–370.

——— (1988a). The Liebig research group—A reappraisal. *Proceedings of the American Philosophical Society* 132:1–66.

——— (1988b). Energy-rich bonds and enzymatic peptide synthesis. In *The Roots of Modern Biochemistry* (H. Kleinkauf, H. von Döhren, and L. Jaenicke, eds.), pp. 165–180. Berlin: de Gruyter.

——— (1990). *Contrasts in Scientific Style: Research Groups in the Chemical and Biochemical Sciences*. Philadelphia: American Philosophical Society.

Fruton, J. S., and Simmonds, S. (1958). *General Biochemistry*. Second ed. New York: Wiley.

Fuerst, J. A. (1982). The role of reductionism in the development of molecular biology: peripheral or central? *Social Studies of Science* 12:241–278.

Fulton, J. F. (1961). *A Bibliography of the Honourable Robert Boyle*. Second ed. Oxford University Press.

Fürth, O. von (1912–1913). *Probleme der physiologischen und pathologischen Chemie*. Leipzig: Vogel.

Futai, M., Noumi, T., and Maeda, M. (1989). ATP synthase (H^+-ATPase): Results by combined biochemical and molecular biological approaches. *Annual Review of Biochemistry* 58:111–136.

Galaty, D. H. (1974). The philosophical basis of mid–nineteenth-century German reductionism. *Journal of the History of Medicine* 29:295–316.

Gale, G. (1984). Science and the philosophers. *Nature* 312:491–495.

Galison, P. (1987). *How Experiments End*. University of Chicago Press.

Gamow, G. (1954). Possible relation between deoxyribonucleic acid and protein structure. *Nature* 173:318.

Garfield, E. (1974). Historiographs, librarianship and the history of science. In *Toward a Theory of Librarianship* (C. H. Rawski, ed.), pp. 380–402. Metuchen, N.J.: Scarecrow Press.

——— (1977). *Essays of an Information Scientist.* Philadelphia: ISI Press.

——— (1979). Trends in biochemical literature. *Trends in Biochemical Sciences* 4:N290-N295.

Garrod, A. E. (1902). The incidence of alkaptonuria: A study in chemical individuality. *Lancet* 2:1616–1620.

——— (1909). *Inborn Errors of Metabolism.* London: Frowde, Hodder, and Stoughton.

Gay-Lussac, J. L. (1809). Mémoire dur la combinaison des substances gazeuses, les avec les autres. *Mémoires de la Société d'Arcueil* 2:207–234.

Gayon, J. (1990). Critics and criticisms of the modern synthesis. *Evolutionary Biology* 24:1–49.

Geertz, C. (1983). *Local Knowledge.* New York: Basic Books.

Gehring, W. J. (1985). The molecular basis of development. *Scientific American* 253(4): 152–162.

Geison, G. L. (1969). The protoplasmic theory of life and the vitalist-mechanist debate. *Isis* 60:273–292.

——— (1971). Ferdinand Julius Cohn. *Dictionary of Scientific Biography* 3:336–341. New York: Charles Scribner's Sons.

——— (1974). Louis Pasteur. *Dictionary of Scientific Biography* 10:350–416. New York: Charles Scribner's Sons.

——— (1978). *Michael Foster and the Cambridge School of Physiology: The Scientific Enterprise in Late Victorian Society.* Princeton University Press.

George, P., and Rutman, R. J. (1960). The "high energy phosphate bond" concept. *Progress in Biophysics and Biophysical Chemistry* 10:1–53.

Gerratt, J. (1987). Modern valence bond theory: Was Kekulé right? *Chemistry in Britain* 23:327–330.

Geyl, P. (1955). *Use and Abuse of History.* New Haven: Yale University Press.

——— (1958). *Debates with Historians.* Cleveland: World Publishing Co.

Giere, R. N. (1989). Scientific rationality as instrumental rationality. *Studies in the History and Philosophy of Science* 20:377–384.

Gilbert, G. N. (1977). Referencing as persuasion. *Social Studies of Science* 7:113–122.

Gillispie, C. C. (1971). Etienne Bonnot, Abbé de Condillac. *Dictionary of Scientific Biography* 3:380–383. New York: Charles Scribner's Sons.

——— (1973). Alexandre Koyré. *Dictionary of Scientific Biography* 7:482–490. New York: Charles Scribner's Sons.

——— (1985). The idea of revolution. *Science* 229:1077–1078.

Glass, B. (1986). Geneticists embattled: Their stand against rampant eugenics and racism in America during the 1920s and 1930s. *Proceedings of the American Philosophical Society* 130:130–154.
Glusker, J. P. (ed.) (1981). *Structural Crystallography in Chemistry and Biology*. London: Hutchinson Ross.
Goldschmidt, R. B. (1938). The theory of the gene. *Scientific Monthly* 46:268–273.
Goldsmith, M. (1980). *Sage: A Life of J. D. Bernal*. London: Hutchinson.
Goldwhite, H. (1975). Clio and chemistry: A divorce has been arranged. *Journal of Chemical Education* 52:645–649.
Golinski, J. (1990). The theory of practice and the practice of theory: Sociological approaches in the history of science. *Isis* 81:492–505.
Golovin, N. E. (1963). The creative person in science. In *Scientific Creativity* (C. W. Taylor and F. Barron, eds.), pp. 7–23. New York: Wiley.
Goodfield, J. (1975). Changing strategies: A comparison of reductionist attitudes in biological and medical research in the nineteenth and twentieth centuries. In *Studies in the Philosophy of Biology* (F. J. Ayala and T. Dobzhansky, eds.), pp. 65–86. London: Macmillan.
Gooding, D., Pinch, T., and Schaffer, S. (1989). *The Uses of Experiment*. Cambridge University Press.
Goodman, D. C. (1969). Problems in crystallography in the early nineteenth century. *Ambix* 16:152–166.
Gortner, R. A. (1929). *Outlines of Biochemistry*. New York: Wiley.
Gotthelf, A. (1987). Aristotle's conception of final causality. In *Philosophical Issues in Aristotle's Biology* (A. Gotthelf and J. G. Lennox, eds.), pp. 204–242. Cambridge University Press.
Gould, S. J. (1976). D'Arcy Thompson and the science of form. In *Topics in the Philosophy of Biology* (M. Grene and E. Mendelsohn, eds.), pp. 66–97. Dordrecht: Reidel.
Graebe, C. (1920). *Geschichte der organischen Chemie*. Berlin: Springer.
Graham, T. (1861). Liquid diffusion applied to analysis. *Philosophical Transactions of the Royal Society of London* 151:183–224.
Green, D. E., Dewan, J. G., and Leloir, L. F. (1937). The beta-hydroxybutyric dehydrogenase of animal tissues. *Biochemical Journal* 31:934–949.
Greenberg, D. S. and Singer, M. F. (1967). The synthesis of DNA: How they spread the good news. *Science* 158:1548–1551.
Grene, M. (1961). The logic of biology. In *The Logic of Personal Knowledge*, pp. 191–205. London: Routledge & Kegan Paul.
——— (1974). *The Understanding of Nature*. Dordrecht: Reidel.
——— (1976). Aristotle in modern biology. In *Topics in the Philosophy of Biology* (M. Grene and E. Mendelsohn, eds.), pp. 3–36. Dordrecht: Reidel.
Grene, M. and Mendelsohn, E. (eds.) (1976). *Topics in the Philosophy of Biology*. Dordrecht: Reidel.

Griffith, B. C., and Mullins, N. C. (1972). Coherent social groups in scientific change. *Science* 177:959–964.

Grimaux, E., and Gerhardt, C. (1900). *Charles Gerhardt, sa Vie, son Oeuvre, sa Correspondance.* Paris: Masson.

Grmek, M. D. (1973). *Raisonnement Expérimental et Recherches Toxicologiques chez Claude Bernard.* Geneva: Droz.

Groenwege, M. P. and Peerdemann, A. F. (1983). Johannes Martin Bijvoet. *Biographical Memoirs of Fellows of the Royal Society* 29:27–41.

Gruber, H. E. (1981). On the relation between "Aha experiences" and the construction of ideas. *History of Science* 19:41–58.

Guerlac, H. (1976). Chemistry as a branch of physics. Laplace's collaboration with Lavoisier. *Historical Studies in the Physical Sciences* 7:193–276.

Guerrier-Takada, C., Gardiner, K., Marsh, T., Pace, N., and Altman, S. (1983). The RNA moiety of ribonuclease P is the catalytic subunit of the enzyme. *Cell* 35:849–857.

Guldberg, C. M., and Waage, P. (1899). Untersuchungen über chemischen Affinitäten. (Translated and edited by R. Abegg.) *Ostwald's Klassiker der exakten Wissenschaften No.104.* Leipzig: Engelmann.

Güttler, H. (1972). Die Begriffe Plasma und Protoplasma: Ihre Entwicklung und Wandlung in der Biologie. *Rete* 1:365–375.

Guyton de Morveau, L. B. (1782). Sur les dénominations chimiques, la nécessité d'en perfectionner le système, et les règles pour parvenir. *Observations sur la Physique, sur l'Histoire Naturelle et sur les Arts et Métiers* 19:370–382.

Haber, E. (1964). Recovery of antigenic specificity after denaturation and complete reduction of disulfides in a papain fragment of antibody. *Proceedings of the National Academy of Sciences* 52:1099–1106.

Hacking, I. (1988). The participant irrealist at large in the laboratory. *British Journal for the Philosophy of Science* 39:277–294.

Haeckel, E. (1866). *Generelle Morphologie der Organismen: Allgemeine Grundzüge der organischen Formen-Wissenschaft, mechanisch begründet durch die von Charles Darwin reformierte Descendenz-Theorie.* Berlin: Reimer.

——— (1917). *Kristallseelen. Studien über das anorganische Leben.* Leipzig: Kroner.

Hagedoorn, A. L. (1911). *Autokatalytic Substances the Determinants for the Inheritable Characters.* Leipzig: Engelmann.

Haldane, J. B. S. (1932). *The Inequality of Man.* London: Chatto & Windus.

——— (1945). A physicist looks at genetics. *Nature* 155:375–376.

Hall, A. R. (1983). On Whiggism. *History of Science* 21:45–59.

Hamburger, V. (1984). Hilde Mangold, co-discoverer of the organizer. *Journal of the History of Biology* 17:1–11.

——— (1988). *The Heritage of Experimental Embryology: Hans Spemann and the Organizer.* New York: Oxford University Press.

Hamilton, W. C. (1970). The revolution in crystallography. *Science* 169:133–141.

Hammarsten, O. (1895). *Lehrbuch der physiologischen Chemie*. Third ed. Wiesbaden: Bergmann.

Häner, R., and Dervan, P. B. (1990). Single-strand DNA triple helix formation. *Biochemistry* 29:9761–9765.

Hanes, C. S. (1937). The action of amylases in relation to the structure of starch and its metabolism in the plant. Part IV. Starch degradation by the component amylases of malt. *New Phytologist* 36:189–239.

Hankins, T. L. (1979). In defense of biography: the use of biography in the history of science. *History of Science* 17:1–16.

Hanson, N. R. (1958). *Patterns of Discovery*. Cambridge University Press.

——— (1971). *What I Do Not Believe, and Other Essays* (S. Toulmin and H. Woolf, eds.). Dordrecht: Reidel.

Haraway, D. (1976). *Crystals, Fabrics and Fields*. New Haven: Yale University Press.

Hargittai, I., and Hargittai, M. (1986). *Symmetry through the Eyes of a Chemist*. Weinheim: VCH Verlagsgesellschaft.

Harrison, E. (1987). Whigs, prigs and historians of science. *Nature* 329:213–214.

Harrison, R. G. (1937). Embryology and its relations. *Science* 85:369–374.

Harrow, B. (1955). *Casimir Funk*. New York: Dodd Mead.

Hartree, E. F. (1976). Ethics for authors: A case history of acrosin. *Perspectives in Biology and Medicine* 20:82–91.

Hastings, A. B. (1989). *Crossing Boundaries*. Grand Rapids: Four Corners Press.

Hauptmann, H. A., and Blessing, R. H. (1987). Fünfundsiebzig Jahre Röntgen-strahlenbeugung und Krystallstrukturanalyse. *Naturwissenschaftliche Rundschau* 40:463–470.

Haynes, R. H. (1989). Genetics and the unity of biology. *Génome* 31:1–7.

Heberer, G. (ed.) (1968). *Der gerechtfertige Haeckel*. Stuttgart: Fischer.

Heider, K., et al. (1919). Dem Andenken an Ernst Haeckel. *Naturwissenschaften* 7:945–971.

Helmholtz, H. (1881). On the modern development of Faraday's conception of electricity. *Journal of the Chemical Society* 39:277–304.

Hempel, C. (1965). *Aspects of Scientific Explanation*. New York: Free Press.

Henderson, L. J. (1928). *Blood: A Study in General Physiology*. New Haven: Yale University Press.

Hendry, J. (1980). Weimar culture and quantum mechanics. *History of Science* 18:115–180.

Herriott, R. M. (1989). Moses Kunitz. *Biographical Memoirs of the National Academy of Sciences* 58:305–317.

Hershey, J. W. B. (1989). Protein phosphorylation controls translation rates. *Journal of Biological Chemistry* 264:20823–20826.

Hess, E. L. (1970). Origins of molecular biology. *Science* 168:664–669.

Hesse, M. B. (1966). *Models and Analogies in Science*. Notre Dame, Ind.: University of Notre Dame Press.

Hickel, E. (1975). Pepsin, Veteran der Enzymchemie. *Naturwissenschaftliche Rundschau* 28:14–18.

——— (1979). Die organische Elementaranalyse. *Pharmazie in unserer Zeit* 8:1–10.

Hiebert, E. N. (1987). The scientist as philosopher of science. *NTM* 24:7–17.

Hill, A. V. (1912). The heat-production of surviving amphibian muscles during rest, activity and rigor. *Journal of Physiology* 44:466–513.

——— (1956). Why biophysics? *Science* 124:1233–1237.

Hinshelwood, C. N. (1953). Autosynthesis. *Journal of the Chemical Society*, pp. 1947–1956.

Hinz, H. J. (ed.) (1986). *Thermodynamic Data for Biochemistry and Biotechnology*. Berlin: Springer.

His, W. (1874). *Unsere Körperform und das physiologische Problem ihrer Entstehung*. Leipzig: Vogel.

Hjelt, E. (1916). *Geschichte der organischen Chemie von ältester Zeit bis zur Gegenwart*. Braunschweig: Vieweg.

Hodgkin, D. C. (1976). Kathleen Lonsdale. *Biographical Memoirs of Fellows of the Royal Society* 21:447–484.

——— (1977). Structures of life. *Chemistry in Britain* 13:138–140.

——— (1980). John Desmond Bernal. *Biographical Memoirs of Fellows of the Royal Society* 26:17–84.

Hoffmann, R. (1988). Under the surface of the chemical article. *Angewandte Chemie (International Edition)* 27:1593–1602.

Hoffmann-Ostenhof, O. (1953). Suggestions for a more rational classification and nomenclature of enzymes. *Advances in Enzymology* 14:219–280.

Hofmann, A. W. (ed.) (1888). *Aus Justus Liebig's und Friedrich Wöhler's Briefwechsel in den Jahren 1829–1873*. Braunschweig: Vieweg.

Hofmeister, F. (1901). *Die chemische Organisation der Zelle*. Braunschweig: Vieweg.

——— (1908). Einiges über die Bedeutung und den Abbau der Eiweisskörper. *Archiv für experimentelle Pathologie und Pharmakologie*, Supplement 273–281.

Hogben, L. (1969). *The Vocabulary of Science*. London: Heinemann.

Hogness, D. S., Cohn, M., and Monod, J. (1955). Studies on the induced synthesis of beta-galactosidase in *Escherichia coli*: The kinetics and mechanism of sulfur incorporation. *Biochimica et Biophysica Acta* 16:99–116.

Holliday, R. (1990). The history of the DNA heteroduplex. *BioEssays* 12:133–142.

Holmes, D. R. (1989). *Stalking the Academic Communist*. Hanover, N.H.: University Press of New England.

Holmes, F. L. (1963). Elementary analysis and the origins of physiological chemistry. *Isis* 54:50–81.

——— (1971). Analysis by fire and solvent extractions: The metamorphosis of a tradition. *Isis* 62:129–148.

——— (1974). *Claude Bernard and Animal Chemistry*. Cambridge, Mass.: Harvard University Press.

——— (1977). Conceptual history. *Studies in the History of Biology* 1:209–218.

——— (1981). The fine structure of scientific creativity. *History of Science* 19:60–70.

——— (1982). Mapping the evolution of biochemistry. *Nature* 300:779–780.

——— (1985). *Lavoisier and the Chemistry of Life*. Madison, Wis.: University of Wisconsin Press.

——— (1986). Patterns of scientific creativity. *Bulletin of the History of Medicine* 60:19–35.

——— (1987a). The intake-output method of quantification in physiology. *Historical Studies in the Physical Sciences* 17:235–270.

——— (1987b). Scientific writing and scientific discovery. *Isis* 78:220–235.

——— (1989a). The complementarity of teaching and research in Liebig's laboratory. *Osiris* [2] 5:121–164.

——— (1989b). Antoine Lavoisier and Hans Krebs: Two styles of scientific creativity. In *Creative People at Work* (D. B. Wallace and H. E. Gruber, eds.), pp. 44–67. New York: Oxford University Press.

——— (1991). *Hans Krebs: The Formation of a Scientific Life*. New York: Oxford University Press.

Homandberg, G. A., Mattis, J. A., and Laskowski, M., Jr. (1978). Synthesis of peptide bonds by proteinases. Addition of solvents shifts peptide bond equilibria toward synthesis. *Biochemistry* 17:5220–5227.

Hooykaas, R. (1952). The species concept in eighteenth-century mineralogy. *Archives Internationales d'Histoire des Sciences* 1:45–55.

——— (1958). The concepts of "individual" and "species" in chemistry. *Centaurus* 5:307–322.

Hopkins, F. G. (1913). The dynamic side of biochemistry. *Nature* 92:213–223.

——— (1933). Some chemical aspects of life. *Science* 78:219–231.

Hoppe-Seyler, F. (1877–1878). *Physiologische Chemie*. Berlin: Hirschwald.

Horder, T. J., Witkowski, J. A., and Wylie, C. C. (eds.) (1985). *A History of Embryology*. Cambridge University Press.

Hoswell, A. E. (1882). Vincenz Kletzinsky. *Berichte der deutschen chemischen Gesellschaft* 15:3310–3315.

Hotchkiss, R. D. (1979). The identification of nucleic acids as genetic determinants. *Annals of the New York Academy of Sciences* 325:321–342.

Hoyningen-Huene, P. (1989). Epistemological reductionism in biology: Intuitions, explications and objections. In *Reductionism and Systems*

Theory (P. Hoyningen-Huene and F. M. Wuketits, eds.), pp. 29–44. Dordrecht: Kluwer.

Huber, R. (1988). Flexibility and rigidity of proteins and protein-complex complexes. *Angewandte Chemie (International Edition)* 27:80–89.

Hughes, A. (1959). *A History of Cytology*. New York: Abelard-Schuman.

Hull, D. L. 1972). Reduction in genetics—biology or philosophy. *Philosophy of Science* 39:491–499.

——— (1974). *Philosophy of Biological Science*. Englewood Cliffs, N.J.: Prentice-Hall.

——— (1979). In defense of presentism. *History and Theory* 18:1–15.

——— (1988). *Science as a Process*. University of Chicago Press.

Hünemörder, C., and Scheele, I. (1977). Das Berufsbild des Biologen im zweiten deutschen Kaiserreich—Anspruch und Wirklichkeit. In *Medizin, Naturwissenschaften und Technik und das zweite Kaiserreich* (G. Mann and R. Winau, eds.), pp. 119–151. Göttingen: Vandenbroeck & Rupprecht.

Hurd, C. D. (1961). The general philosophy of organic nomenclature. *Journal of Chemical Education* 38:43–47.

Huxley, A. (1963). *Literature and Science*. New York: Harper & Row.

Huxley, T. H. (1869). On the physical basis of life. *The Fortnightly Review*, n.s.5:129–145.

Iterson, G. van (1940). *Martinus Willem Beijerinck, His Life and Work*. The Hague: Nijhoff.

Jacob, F. (1970). *La Logique du Vivant*. Paris: Gallimard.

——— (1973). *The Logic of Life* (translated by B. E. Spillmann). New York: Pantheon.

——— (1982). *The Possible and the Actual*. Seattle: University of Washington Press.

——— (1987). *La Statue Intérieure*. Paris: Seuil.

Jacob, F., and Monod, J. (1961). Genetic regulatory mechanisms in the synthesis of proteins. *Journal of Molecular Biology* 3:318–356.

Jacques, J. (1987). *Berthelot: Autopsie d'un Mythe*. Paris: Belon.

James, F. A. J. L. (1985). The creation of a Victorian myth: The historiography of spectroscopy. *History of Science* 23:1–24.

James, M. N. G., and Sielecki, A. R. (1983). Structure and refinement of penicillopepsin at 1.8 A resolution. *Journal of Molecular Biology* 163:299–361.

Janin, J. (1990). Errors in three dimensions. *Biochimie* 72:705–709.

Jarausch, K. H. (ed.) (1983). *The Transformation of Higher Learning 1860–1930*. Stuttgart: Klett-Cotta.

Jeffrey, C. (1973). *Biological Nomenclature*. London: Edward Arnold.

Jencks, W. P. (1975). Binding energy, specificity and enzymic catalysis—The Circe effect. *Advances in Enzymology* 43:219–410.

Johnson, J. (1985a). Academic chemistry in imperial Germany. *Isis* 76:500–524.

——— (1985b). Academic self-regulation and the chemical profession in Germany. *Minerva* 23:241–271.

Judson, H. F. (1979). *The Eighth Day of Creation: The Makers of the Revolution in Biology.* New York: Simon and Schuster.

Kalckar, H. M. (1941). The nature of energetic coupling in biochemical syntheses. *Chemical Reviews* 28:71–178.

Kant, I. (1928). *Critique of Teleological Judgement* (translated by J. C. Meredith). Oxford University Press.

Kantrowitz, E. R., and Lipscomb, W. N. (1990). *Escherichia coli* aspartate transcarbamoylase: The molecular basis for a concerted allosteric transition. *Trends in Biochemical Sciences* 15:53–59.

Karlson, P. (1986). Wie und warum entstehen wissenschaftliche Irrtümer? *Naturwissenschaftliche Rundschau* 39:380–389.

Katz, M. J. (1987). Are there biological impossibilities? In *No Way, the Nature of the Impossible* (P. J. Davis and D. Park, eds.), pp. 11–43. New York: Freeman.

Kautzsch. K. (1906). Emil Fischers Forschungen auf dem Gebiete der Eiweisschemie. *Die Umschau* 10:129–131.

Kay, L. E. (1985). Conceptual models and analytical tools: The biology of Max Delbrück. *Journal of the History of Biology* 18:207–246.

——— (1990). Review of A. Serafini, *Linus Pauling: A Man and his Science.* *Bulletin of the History of Medicine* 64:496–498.

Keilin, D. (1966). *The History of Cell Respiration and Cytochrome.* Cambridge University Press.

Kendrew, J. C. (1968). Information and conformation in biology. In *Structural Chemistry and Molecular Biology* (A. Rich and N. Davidson, eds.), pp. 187–197. San Francisco: Freeman.

Kendrew, J. C., Bodo, G., Dintzis, H. M., Parrish, R. G., and Wyckoff, H. (1958). A three-dimensional model of the myoglobin molecule obtained by X-ray analysis. *Nature* 181:662–666.

Kenyon, R. L. (1966). Problems of anonymity. *Chemical and Engineering News* 44(45): 5.

Keyser, B. W. (1990). Between science and craft: The case of Berthollet and dyeing. *Annals of Science* 47:213–260.

Klein, D. J. and Trinajstic, N. (1990). Valence-bond theory and chemical structure. *Journal of Chemical Education* 67:633–637.

Klein, M. (1936). *Histoire des Origines de la Théorie Cellulaire.* Paris: Hermann.

Kletzinsky, V. (1858). *Compendium der Biochemie.* Vienna: Braumüller.

Klosterman, L. J. (1985). A research school of chemistry in the nineteenth century: Jean Baptiste Dumas and his research students. *Annals of Science* 42:1–80.

Klotz, I. M. (1986). *Introduction to Biomolecular Energetics.* Orlando, Fla.: Academic Press.

Kluyver, A. J., and Donker, H. J. L. (1926). Die Einheit in der Biochemie. *Chemie der Zelle und Gewebe* 13:134–190.

Kluyver, A. J., and Van Niel, C. B. (1956). *The Microbe's Contribution to Biology*. Cambridge, Mass.: Harvard University Press.

Knowles, J. R. (1965). Enzyme specificity: alpha-chymotrypsin. *Journal of Theoretical Biology* 9:213–228.

——— (1991). Enzyme catalysis: not different, just better. *Nature* 350:121–124

Kohler, R. E. (1971). The background to Eduard Buchner's discovery of cell-free fermentation. *Journal of the History of Biology* 4:35–61.

——— (1972). The reception of Eduard Buchner's discovery of cell-free fermentation. *Journal of the History of Biology* 5:327–353.

——— (1975). The history of biochemistry: A survey. *Journal of the History of Biology* 8:275–318.

——— (1976). The management of science: The experience of Warren Weaver and the Rockefeller Foundation programme in molecular biology. *Minerva* 14:279–306.

——— (1977). Warren Weaver and the Rockefeller Foundation program in molecular biology: A case study in the management of science. In *The Scientist in the American Context: New Perspectives* (N. Reingold, ed.), pp. 249–293. Washington, D.C.: Smithsonian Institution.

——— (1978). Walter Fletcher, F. G. Hopkins, and the Dunn Institute of Biochemistry: A case history of the patronage of science. *Isis* 69:331–355.

——— (1982). *From Medical Chemistry to Biochemistry*. Cambridge University Press.

——— (1985a). Bacterial physiology: The medical context. *Bulletin of the History of Medicine* 59:54–74.

——— (1985b). Innovation in normal science: Bacterial physiology. *Isis* 76:162–181.

——— (1991). *Partners in Science. Foundations and Natural Scientists, 1900–1945*. University of Chicago Press.

Koltzov, N. K. (1928). Physikalisch-chemische Grundlage der Morphologie. *Biologisches Zentralblatt* 48:345–369.

Kornberg, A. (1987). The two cultures: Chemistry and biology. *Biochemistry* 26:6888–6891.

——— (1988). DNA replication. *Journal of Biological Chemistry* 263:1–4.

——— (1989). *For the Love of Enzymes*. Cambridge, Mass.: Harvard University Press.

Körner, S. (1959). *Conceptual Thinking*. New York: Dover.

——— (1970). *Categorial Frameworks*. Oxford: Blackwell.

——— (1971). *Abstraction in Science and Morals*. Cambridge University Press.

——— (1976). *Fundamental Questions of Philosophy*. Fourth ed. Brighton, Sussex: Harvester Press.

——— (1986). On scientific information, explanation and progress. In *Logic, Methodology, and Philosophy of Science VII* (R. Barcan Marcus et al., eds.), pp. 1–15. Amsterdam: North Holland.

——— (1990). *Kant*. Harmondsworth: Penguin.

Korr, I. M. (1939). Oxidation-reductions in heterogeneous systems. *Cold Spring Harbor Symposia on Quantitative Biology* 7:74–93.

Koshland, D. E. (1958). Application of a theory of enzyme specificity to protein synthesis. *Proceedings of the National Academy of Sciences* 44:98–104.

Koshland, D. E., Nemethy, G., and Filmer, D. (1966). Comparison of binding data and theoretical models in proteins containing subunits. *Biochemistry* 5:365–385.

Koyré, E. (1956). Les origines de la science moderne. *Diogène* 16(4):14–42.

Krafft, C. F. (1938). *The Chemical Organization of Living Matter* Second ed. Washington, D.C.: privately printed.

Kragh, H. (1987). *An Introduction to the Historiography of Science*. Cambridge University Press.

Kraut, J. (1988). How do enzymes work? *Science* 242:533–540.

Krebs, H. A. (1971). How the whole becomes more than the sum of its parts. *Perspectives in Biology and Medicine* 14:448–457.

——— (1976). Comments on the productivity of scientists. In *Reflections on Biochemistry* (A. Kornberg et al., eds.), pp. 415–421. Oxford: Pergamon.

——— (1981). *Reminiscences and Reflections*. Oxford University Press.

Krebs, H. A., and Henseleit, K. (1932). Untersuchungen über die Harnstoffbildung im Tierkörper. *Zeitschrift für physiologische Chemie* 210:33–66.

Krebs, H. A., and Kornberg, H. L. (1957). A survey of the energy transformations in living matter. *Ergebnisse der Physiologie* 49:212–298.

Krebs, H. A., and Shelley, J. H. (eds.) (1974). *The Creative Process in Science and Medicine*. Amsterdam: Excerpta Medica.

Kreil, G. (1990). Processing of precursors by dipeptidylaminopeptidases: A case of molecular ticketing. *Trends in Biochemical Sciences* 15:23–26.

Kuhn, R. (1935). Sur les flavines. *Bulletin de la Société de Chimie Biologique* 17:905–926.

Kuhn, T. S. (1977). *The Essential Tension*. University of Chicago Press.

——— (1983). Commensurability, comparability, communicability. *PSA 1982* 2:669–716.

——— (1989). Possible worlds in history of science. In *Possible Worlds in Humanities, Arts, and Sciences* (S. Allen, ed.), pp. 9–32, 49–50. Berlin: de Gruyter.

Kühne, W. (1866). *Lehrbuch der physiologischen Chemie*. Leipzig: Engelmann.

Kulkarni, D., and Simon, H. A. (1988). The processes of scientific discovery: The strategy of experimentation. *Cognitive Science* 12:139–175

Kümmel, W. F. (1989). Analogie in den Wissenschaften: Einführung in das Thema. *Berichte zur Wissenschaftsgeschichte* 12:1–6.

Landsteiner, K. (1946). *The Specificity of Serological Reactions*. Cambridge, Mass.: Harvard University Press.

Langley, P., Simon, H. A., Bradshaw, G. L., and Zytkow, J. M. (1987). *Scientific Discovery*. Cambridge, Mass.: The MIT Press.

Langmuir, I. (1939). The structure of proteins. *Proceedings of the Physical Society* 51:592–612.

Latour, B., and Woolgar, S. (1986). *Laboratory Life: The Construction of Scientific Facts*. Princeton University Press.

Laudan, L. (1968). Theories of scientific method from Plato to Mach. *History of Science* 7:1–63.

Laudan, R. (1983). Redefinitions of a discipline: Histories of geology and geological history. In *Functions and Uses of Disciplinary Histories* (L. Graham, W. Lepienes, and P. Weingart, eds.), pp. 79–103. Dordrecht: Reidel.

——— (1987). *From Mineralogy to Geology: The Foundations of a Science, 1650–1830*. University of Chicago Press.

——— (1989). Individuals, species and the development of mineralogy and geology. In *What the Philosophy of Biology Really Is* (M. Ruse, ed.), pp. 221–233. Dordrecht: Kluwer.

Lavoisier, A. L. (1790). *Elements of Chemistry* (translated by R. Kerr). Edinburgh: Creech.

Law, J. (1973). The development of specialties in science: The case of X-ray protein crystallography. *Science Studies* 3:275–303.

Lea, D. E., Haines, R. B., and Coulson, C. A. (1936). The mechanism of the bactericidal action of radioactive radiations. *Proceedings of the Royal Society* B120:47–76.

Leavis, F. R. (1962). *Two Cultures? The Significance of C. P. Snow*. London: Chatto and Windus.

Lederberg, J. (1987). Genetic recombination of bacteria: A discovery account. *Annual Review of Genetics* 21:23–46.

Lederberg, J., and Tatum, E. L. (1946). Gene recombination in *Escherichia coli*. *Nature* 158:558.

Lehman, I. R., Bessman, M. J., Simms, E. S., and Kornberg, A. (1958). Enzymatic synthesis of deoxyribonucleic acid. *Journal of Biological Chemistry* 233:163–177.

Lehman, I. R., and Kaguni, L. S. (1989). DNA polymerase alpha. *Journal of Biological Chemistry* 264:4265–4268.

Lehmann, C. G. (1853). *Lehrbuch der physiologischen Chemie*. Second ed. Leipzig: Engelmann.

Lehmann, O. (1907). *Die scheinbar lebenden Kristalle*. Munich: Schreiber.

——— (1911). *Die neue Welt der flüssigen Kristalle und deren Bedeutung für Physik, Chemie, Technik und Biologie*. Leipzig: Akademische Verlagsgesellschaft.

Lehn, J. M. (1988). Supramolecular chemistry—Scope and perspectives. Molecules, supermolecules and molecular devices. *Angewandte Chemie (International Edition)* 27:90–112.

Lehninger, A. L. (1975). *Biochemistry.* Second ed. New York: Worth.

Lejeune, P. (1989). *On Autobiography.* Minneapolis: University of Minnesota Press.

Leloir, L. F. (1971). Two decades of research on the biosynthesis of saccharides. *Science* 172:1299–1302.

Lerner, R. A., and Benkovic, S. J. (1988). Principles of antibody catalysis. *BioEssays* 9:107–112.

Lerner, R. A., Benkovic, S. J., and Schultz, P. G. (1991). At the crossroads of chemistry and immunology: Catalytic antibodies. *Science* 252:659–667.

Levere, T. H. (1971). *Affinity and Matter.* Oxford University Press.

Lewis, G. N., and Randall, M. (1923). *Thermodynamics and the Free Energy of Chemical Substances.* New York: McGraw-Hill.

Lewis, H. D. (ed.) (1963). *Clarity Is Not Enough.* London: Allen and Unwin.

Lewis, I. M. (1934). Bacterial variation with special reference to some mutable strains of colon bacteria in synthetic media. *Journal of Bacteriology* 28:619–638.

Lieben, F. (1935). *Geschichte der physiologischen Chemie.* Leipzig: Deuticke.

Liebig, J. (1842). *Die organische Chemie in ihrer Anwendung auf Physiologie und Pathologie.* Braunschweig: Vieweg.

——— (1874). *Reden und Abhandlungen.* Leipzig: Winter.

Liegener, C., and Del Re, G. (1987). Chemistry vs. physics, the reductionist myth, and the unity of science. *Zeitschrift für allgemeine Wissenschaftsgeschichte* 18:165–174.

Lindauer, M. W. (1962). The evolution of the concept of chemical equilibrium from 1775 to 1923. *Journal of Chemical Education* 39:384–390.

Lindeboom, G. A. (1957). From the history of the concept of specificity. *Janus* 46:12–24.

Linderstrøm-Lang, K., Hotchkiss, R. D., and Johansen, G. (1938). Peptide bonds in proteins. *Nature* 142:996–997.

Lipmann, F. (1941). Metabolic generation and utilization of phosphate bond energy. *Advances in Enzymology* 1:99–162.

——— (1950). Biosynthetic mechanisms. *Harvey Lectures* 44:99–123.

——— (1954). On the mechanism of some ATP-linked reactions and certain aspects of protein synthesis. In *Mechanism of Enzyme Action* (W. D. McElroy and B. Glass, eds.), pp. 599–604. Baltimore: The John Hopkins University Press.

——— (1971). *Wanderings of a Biochemist.* New York: Wiley.

Lloyd, G. E. R. (1987). Empirical research in Aristotle's biology. In *Philosophical Issues in Aristotle's Biology* (A. Gotthelf and J. G. Lennox, eds.), pp. 53–63. Cambridge University Press.

Lodish, H. F. (1982). Validity of scientific data—The responsibility of the principal investigator. *Trends in Biochemical Sciences* 7:86–87.

Loeb, J. (1916). *The Organism as a Whole*. New York: Putnam.

Loew, O. (1896). *The Energy of Living Protoplasm*. London: Kegan Paul, Trench, Trübner and Co.

Lorch, J. (1974). The charisma of crystals in biology. In *The Interaction between Science and Philosophy* (Y. Elkana, ed.), pp. 445–461. Atlantic Highlands, N.J.: Humanities Press.

Lovejoy, A. O. (1936). *The Great Chain of Being*. Cambridge, Mass.: Harvard University Press.

Löwy, I. (1988). Immunology and literature in the early twentieth century: *Arrowsmith* and the *Doctor's Dilemma*. *Medical History* 32:314–332.

Ludmerer, K. M. (1972). *Genetics and American Society*. Baltimore: The Johns Hopkins University Press.

Luisi, P. L., and Thomas, R. M. (1990). The pictographic molecular paradigm. *Naturwissenschaften* 77:67–74.

Lundgren, A. (1990). The changing role of numbers in 18th-century chemistry. In *The Quantifying Spirit in the 18th Century* (T. Frangsmyr, J. L. Heilbron, and R. E. Rider, eds.), pp. 245–266. Berkeley: University of California Press.

Luria, S. E. (1939). Action des radiations sur le *Bacterium coli*. *Comptes Rendus de l'Académie des Sciences* 209:604–606.

——— (1970). Molecular biology: Past, present and future. *BioScience* 20:1289–1293,1296.

——— (1984). *A Slot Machine, a Broken Test Tube*. New York: Harper & Row.

Luria, S. E., and Delbrück, M. (1943). Mutations of bacteria from virus sensitivity to virus resistance. *Genetics* 28:491–511.

Lusk, G. (1928). *Elements of the Science of Nutrition*. Fourth ed. Philadelphia: Saunders.

Lwoff, A. (1962). *Biological Order*. Cambridge, Mass.: The MIT Press.

——— (1977). Jacques Lucien Monod. *Biographical Memoirs of Fellows of the Royal Society* 23:385–412.

Lwoff, A., and Ullmann, A. (eds.) (1979). *Origins of Molecular Biology: A Tribute to Jacques Monod*. New York: Academic Press.

McCarty, M. (1985). *The Transforming Principle*. New York: Norton.

McGlashan, M. L. (1966). The use and misuse of the laws of thermodynamics. *Journal of Chemical Education* 43:226–232.

McIlwain, H. (1990). Biochemistry and neurochemistry in the 1800s: Their origins in comparative animal biochemistry. *Essays in Biochemistry* 25:197–224

McKusick, V. A. (1960). Walter S. Sutton and the physical basis of Mendelism. *Bulletin of the History of Medicine* 34:487–497.

McLaughlin, P. (1989). *Kants Kritik der teleologischen Urteilskraft*. Bonn: Bouvier.

MacRoberts, M. H. and MacRoberts, B. R. (1984). The negational reference: or the art of dissembling. *Social Studies of Science* 14:91–94.

——— (1989). Citation analysis and the science policy area. *Trends in Biochemical Sciences* 14:8, 10, 12.

Maddox, J. (1964). Is the literature worth keeping? *Graduate Journal* 6:169–178.

——— (1976). Biochemistry and Babel. *Trends in Biochemical Sciences* 1:N3.

——— (1989). Where next with peer-review? *Nature* 339:11.

Maienschein, J. (1983). Experimental biology in transition: Harrison's embryology 1895–1910. *Studies in History of Biology* 6:107–127.

——— (1985). History of Biology. *Osiris* [2] 1:147–162.

Mandelstam, J. (1958). Turnover of protein in growing and non-growing populations of *Escherichia coli*. *Biochemical Journal* 69:110–119.

Mann, G. (1906). *Chemistry of the Proteids*. London: Macmillan.

Manwaring, W. H. (1930). Renaissance of pre-Ehrlich immunology. *Journal of Immunology* 19:155–162.

Marchionni, M., and Gilbert, W. (1986). The triose phosphate gene from maize: Introns antedate the plant-animal divergence. *Cell* 46:133–141.

Markert, C. L., and Whitt, G. S. (1968). Molecular varieties of enzymes. *Experientia* 24:977–991.

Markert, C. L., Shaklee, J. B., and Whitt, G. S. (1976). Evolution of a gene. *Science* 189:102–114.

Marwick, A. (1989). *The Nature of History*. Third ed. London: Macmillan.

Marx, J. (1974). L'art d'observer au XVIIIe siècle: Jean Senebier et Charles Bonnet. *Janus* 61:201–220.

Maupertius, P. L. (1756). *Oeuvres (Nouvelle Édition)*. Lyon: Bruyset.

Mauskopf, S. (1976). Crystals and compounds: Molecular structure and composition in nineteenth-century France. *Transactions of the American Philosophical Society* 66:1–82.

Maxam, A. M. and Gilbert, W. (1977). A new method for sequencing DNA. *Proceedings of the National Academy of Sciences* 74:560–564.

Mayr, E. (1961). Cause and effect in biology. *Science* 134:1501–1506.

——— (1968). The role of systematics in biology. *Science* 159:595–599.

——— (1976). *Evolution and the Diversity of Life*. Cambridge, Mass.: Harvard University Press.

——— (1982). *The Growth of Biological Thought*. Cambridge, Mass.: Harvard University Press.

——— (1988). *Toward a New Philosophy of Biology*. Cambridge, Mass.: Harvard University Press.

——— (1989). Attaching names to objects. In *What the Philosophy of Biology Is* (M. Ruse, ed.), pp. 235–243. Dordrecht: Kluwer.

——— (1990). When is historiography whiggish? *Journal of the History of Ideas* 51:301–309

Mayr, E., and Provine, W. B. (eds.) (1980). *The Evolutionary Synthesis.* Cambridge, Mass.: Harvard University Press.

Mayr, E., and Short, L. L. (1970). *Species Taxa of North American Birds.* Cambridge, Mass.: Nuttall Ornithological Club.

Mazumdar, P. M. H. (1974). The antigen-antibody reaction and the physics and chemistry of life. *Bulletin of the History of Medicine* 48:1–21.

Mazur, A. (1989). Allegations of dishonesty in research and their treatment by American universities. *Minerva* 27:177–194.

Medawar, P. B. (1944). The behaviour and fate of skin autographs and skin homografts in rabbits. *Journal of Anatomy* 78:176–199.

——— (1945a). A second study of the behaviour of skin homografts in rabbits. *Journal of Anatomy* 79:117–156.

——— (1945b). Size, shape, and age. In *Essays on Growth and Form Presented to D'Arcy Wentworth Thompson* (W. E. LeGros Clark and P. B. Medawar, eds.), pp. 157–187. Oxford University Press.

——— (1963). Is the scientific paper a fraud? *The Listener*, 12 September, p. 377.

——— (1967). *The Art of the Soluble.* London: Methuen.

——— (1968). Lucky Jim. *New York Review of Books*, 28 March, pp. 3–5.

——— (1969). *Induction and Intuition in Scientific Thought.* Philadelphia: American Philosophical Society.

——— (1979). *Advice to a Young Scientist.* New York: Harper & Row.

——— (1986). *Memoir of a Thinking Radish.* Oxford University Press.

Medawar, P. B., and Medawar, J. S. (1983). *Aristotle to Zoos.* Cambridge, Mass.: Harvard University Press.

Merton, R. K. (1957). *Social Theory and Social Structure.* Glencoe, Ill.: Free Press.

——— (1961). Singletons and multiples in scientific discovery: A chapter in the sociology of science. *Proceedings of the American Philosophical Society* 105:470–486.

——— (1963). The ambivalence of scientists. *Bulletin of the Johns Hopkins Hospital* 112:77–97.

——— (1965). *On the Shoulders of Giants.* New York: Free Press.

——— (1968a). The Matthew effect in science. *Science* 159:56–53.

——— (1968b). Making it scientifically. *The New York Times Book Review*, 25 February, pp. 41–43, 45.

——— (1973). *The Sociology of Science.* University of Chicago Press.

——— (1988). The Matthew effect in science II. *Isis* 79:606–623.

Meselson, M., and Stahl, F. W. (1958). The replication of DNA in *Escherichia coli*. *Proceedings of the National Academy of Sciences* 44:671–682.

Metchnikoff, E. (1901). *L'Immunité dans les Maladies Infectueuses.* Paris: Masson.

Metzger, H. (1918). *La Genèse de la Science des Cristaux.* Paris: Alcan.

Meyerhof, O. (1926). Thermodynamik des Lebensprozesses. In *Handbuch der Physik*, vol. 11 (F. Hennig, ed.), pp. 239–271. Berlin: Springer.

——— (1930). *Die chemische Vorgänge im Muskel*. Berlin: Springer.
Miescher, F. (1897). *Die histochemische und physiologische Arbeiten* (W. His, ed.). Leipzig: Vogel.
Millar, M., and Millar, I. T. (1988). Chemists as autobiographers. *Journal of Chemical Education* 65:847–853.
Miller, J. G. (1978). *Living Systems*. New York: McGraw-Hill.
Miller, R. W. (1987). *Fact and Method*. Princeton University Press.
Miller, S. L., and Orgel, L. E. (1974). *The Origins of Life on Earth*. Englewood Cliffs, N.J.: Prentice-Hall.
Mills, H. W. (1939). Arthur Hutchinson. *Journal of the Chemical Society*, pp. 210–213.
Mitchell, P. (1979). Keilin's respiratory chain concept and its chemi-osmotic consequences. *Science* 206: 1148–1159.
Mitchison, A. N. (1990). Peter Brian Medawar. *Biographical Memoirs of Fellows of the Royal Society* 35: 283–301.
Monod, J. (1942). *La Croissance des Cultures Bactériennes*. Paris: Hermann.
——— (1947). The phenomenon of enzymatic adaptation and its bearings on problems of genetics and cellular differentiation. *Growth Symposium* 11:223–289.
——— (1970). *Le Hasard et la Nécessité*. Paris: Seuil.
——— (1971). Du microbe à l'homme. In *Of Microbes and Life* (J. Monod and E. Borek, eds.), pp. 1–9. New York: Columbia University Press.
Monod, J., Changeux, J. P., and Jacob, F. (1963). Allosteric proteins and cellular control systems. *Journal of Molecular Biology* 6:306–329.
Monod, J., Wyman, J., and Changeux, J. P. (1965). On the nature of allosteric transitions: A plausible model. *Journal of Molecular Biology* 12:88–118.
Moore, W. (1989). *Schrödinger, Life and Thought*. Cambridge University Press.
Morgan, T. H. (1910). Sex limited inheritance in Drosophila. *Science* 32:120–122.
——— (1926). *The Theory of the Gene*. New Haven: Yale University Press.
Morgan, T. H., Sturtevant, A. H., Muller, H. J., and Bridges, C. B. (1915). *The Mechanism of Mendelian Heredity*. New York: Holt.
Morrell, J. B. (1972). The chemist breeders: The research schools of Liebig and Thomas Thomson. *Ambix* 19:1–46.
Muller, H. J. (1922). Variation due to change in the individual gene. *American Naturalist* 56:32–50.
——— (1927). Artificial transmutation of the gene. *Science* 66:84–87.
——— (1929). The gene as the basis of life. *Proceedings of the International Congress of Plant Sciences* 1:897–921.
——— (1943). Edmund B. Wilson—An appreciation. *American Naturalist* 77:5–37,142–172.
Mullins, N. C. (1972). The development of a scientific specialty: The Phage Group and the origins of molecular biology. *Minerva* 10:51–82.

Multhauf, R. P. (1966). *The Origins of Chemistry.* London: Oldbourne.

Mutter, M. (1985). The construction of new proteins and enzymes—A prospect for the future? *Angewandte Chemie (International Edition)* 24:639–653.

Nagel, E. (1961). *The Structure of Science.* New York: Harcourt Brace.

Nägeli, C. (1879). Theorie von Gärung. *Abhandlungen der königlichen Akademie der Wissenschaften München* 13(2): 77–205.

——— (1884). *Mechanisch-physiologische Theorie der Abstammungslehre.* Leipzig: Oldenbourg.

Needham, A. E. (1965). *The Uniqueness of Biological Materials.* Oxford: Pergamon.

Needham, D. M. (1971). *Machina Carnis: The Biochemistry of Muscular Contraction and its Historical Development.* Cambridge University Press.

Needham, J. (1929). *The Sceptical Biologist.* London: Chatto & Windus.

——— (1936). *Order and Life.* New Haven: Yale University Press.

——— (1959). *A History of Embryology.* Second ed. Cambridge University Press.

Neuberger, A., and Brocklehurst, K. (eds.) (1987). *Hydrolytic Enzymes.* Amsterdam: Elsevier.

Neumann, J. (1989). Der historisch-soziale Ansatz medizinischer Wissenschaftstheorie von Ludwik Fleck. *Sudhoffs Archiv* 73:12–25.

Neumeister, R. (1903). *Betrachtungen über das Wesen der Lebenserscheinungen.* Jena: Fischer.

Newcomb, S. (1990). Contributions of British experimentalists to the discipline of geology 1780–1820. *Proceedings of the American Philosophical Society* 134:161–225.

Nickles, T. (1987). Methodology, heuristics and rationality. In *Rational Changes in Science* (J. C. Pitt and M. Pera, eds.), pp. 103–132. Dordrecht: Reidel.

Nisinoff, A. (1985). *Introduction to Molecular Immunology.* Second ed. Sunderland, Mass.: Sinauer.

Northrop, J. H. (1961). Biochemists, biologists, and William of Occam. *Annual Review of Biochemistry* 30:1–10.

Northrop, J. H., Kunitz, M., and Herriott, R. M. (1948). *Crystalline Enzymes.* Second ed. New York: Columbia University Press.

Nossal, G. J. V. (1986). Turning points in cellular immunology: The skein untangled through a global invisible college. *Perspectives in Biology and Medicine* 29:S166-S177.

Novikoff, A. B. (1945). The concept of integrative levels and biology. *Science* 101:209–215.

Olby, R. C. (1966). *Origins of Mendelism.* London: Constable.

——— (1974). *The Path to the Double Helix.* Seattle: University of Washington Press.

——— (1985). The 'mad pursuit': X-ray crystallographers' search for the

structure of hemoglobin. *History and Philosophy of the Life Sciences* 7:171–193.
——— (1986). Biochemical origins of molecular biology: A discussion. *Trends in Biochemical Sciences* 11:303–305.
——— (1990). The molecular revolution in biology. In *Companion to the History of Modern Science* (R. C. Olby et al., eds.), pp. 503–520. London: Routledge.
Olby, R. C., Cantor, G. N., Christie, J. R. R., and Hodge, M. J. S. (1990). *Companion to the History of Modern Science*. London: Routledge.
Olesko, K. M. (1988). On institutes, investigations, and scientific training. In *The Investigative Enterprise in Experimental Physiology in Nineteenth-Century Medicine* (W. Coleman and F. L. Holmes, eds.), pp. 295–332. Berkeley: University of California Press.
Oppenheimer, J. M. (1967). *Essays in the History of Embryology and Biology*. Cambridge, Mass.: The MIT Press.
——— (1972). Ross Granville Harrison. *Dictionary of Scientific Biography* 6:131–135. New York: Charles Scribner's Sons.
Orel, V. (1984). *Mendel*. Oxford University Press.
Ortony, A. (ed.) (1979). *Metaphor and Thought*. Cambridge University Press.
Ostrowski, P. F. (1986). Who discovered vitamins? *The Polish Review* 31:171–183.
Ostwald, W. (1900). Ueber Oxydationen mittels freien Sauerstoffs. *Zeitschrift für physikalische Chemie* 34:248–252.
Page, M. I. (1979). Entropy, binding energy, and enzymic catalysis. *Angewandte Chemie (International Edition)* 16:449–459.
Pagel, W. (1982a). *Paracelsus*. Second ed. Basle: Karger.
——— (1982b). *Johan Baptist van Helmont*. Cambridge University Press.
Pantin, C. F. A. (1968). *The Relations between the Sciences*. Cambridge University Press.
Parascandola, J. (1974). The controversy over structure-function relationships in the early twentieth century. *Pharmacy in History* 6:54–63.
——— (1975). The evolution of stereochemical concepts in pharmacology. In *Van't Hoff—Le Bel Centennial* (O. B. Ramsay, ed.), pp. 143–158. Washington, D.C.: American Chemical Society.
Parascandola, J. (ed.) (1980). The History of Antibiotics. Madison, Wis.: American Institute of the History of Pharmacy.
Parnas, J., Ostern, P., and Mann, T. (1934). Ueber die Verkettung der chemischen Vorgänge im Muskel. *Biochemische Zeitschrift* 272:64–70.
Partington, J. R. (1961–1970). *A History of Chemistry*. Vol. 1, pt. 1 (1970); vol. 2 (1961); vol. 3 (1952); vol. 4 (1964). London: Macmillan.
Pasteur, L. (1879). *Examen Critique d'un Écrit Posthume de Claude Bernard sur la Fermentation*. Paris: Gauthier-Villars.
Paul, H. W. (1976). Scholarship and ideology: The chair of the general history of science at the Collège de France 1892–1913. *Isis* 67:376–397.

Pauling, L. (1940). A theory of the structure and process of formation of antibodies. *Journal of the American Chemical Society* 62:2643–2657.

——— (1948). Nature of forces between large molecules of biological molecules of biological interest. *Nature* 161:707–709.

——— (1955). The stochastic method and the structure of proteins. *American Scientist* 43:285–297.

——— (1970). Fifty years of progress in structural chemistry and molecular biology. *Daedalus* 99:988–1014.

——— (1971). Roscoe Gilkey Dickinson. *Dictionary of Scientific Biography* 4:82. New York: Charles Scribner's Sons.

——— (1974). Molecular basis of biological specificity. *Nature:* 248:769–771.

Pauling, L., and Campbell, D. H. (1942). The manufacture of antibodies in vitro. *Journal of Experimental Medicine* 76:211–230; *Science* 95:440–441.

Pauling, L., Corey, R. B., and Branson, H. R. (1951). The structure of proteins. Two hydrogen-bond and helical configurations of the polypeptide chain. *Proceedings of the National Academy of Sciences* 37:205–211.

Pauling, L., and Niemann, C. (1939). The structure of proteins. *Journal of the American Chemical Society* 61:1860–1867.

Peacocke, A. R. (1983). *An Introduction to the Physical Chemistry of Biological Organization.* Oxford University Press.

Pelikan, J. (1984). Scholarship: a sacred vocation. *Scholarly Publishing* 16:3–22.

Pellegrin, P. (1987). Logical difference and biological difference: The unity of Aristotle's thought. In *Philosophical Issues in Aristotle's Biology* (A. Gotthelf and J. E. Lennox, eds.), pp. 313–338. Cambridge University Press.

Pérez-Ramos, A. (1988). *Francis Bacon's Idea of Science and the Maker's Knowledge Tradition.* Oxford University Press.

Perrin, C. E. (1987). Revolution or reform: The chemical revolution and eighteenth-century concepts of scientific change. *History of Science* 25:395–423.

Perutz, M. F. (1970). Stereochemistry of cooperative effects in haemoglobin. *Nature* 228:726–739.

——— (1980). Origins of molecular biology. *New Scientist* 85:326–329.

——— (1985). Early days of protein crystallography. *Methods in Enzymology* 114:3–18.

——— (1986a). A new view of Darwinism. *New Scientist* 112(2 October):35–38.

——— (1986b). Keilin and the Molteno. *Cambridge Review* 107:152–156.

——— (1987). Physics and the riddle of life. *Nature* 326:555–558.

——— (1988). Reply, from Perutz, on reductionism. *Trends in Biochemical Sciences* 13:206.

——— (1989a). Seeking to reveal the face of nature. *Times Literary Supplement*, 6–12 January, p. 8.

——— (1989b). *Is Science Necessary?* New York: Dutton.

——— (1990). *Mechanisms of Cooperativity and Allosteric Regulation in Proteins*. Cambridge University Press.

Peters, R. A. (1930). Surface structure in the integration of cell activity. *Transaction of the Faraday Society* 26:797–807.

Pflüger, E. (1875). Beiträge zur Lehre von der Respiration. I. Ueber die physiologische Verbrennung in den lebendigen Organismen. *Archiv für die gesamte Physiologie des Menschen und der Tiere* 10:251–369, 641–644.

——— (1878). Ueber Wärme und Oxydation der lebendigen Materie. *Archiv für die gesamte Physiologie des Menschen und der Tiere* 18:247–380.

Phillips, J. P. (1966). Liebig and Kolbe, critical editors. *Chymia* 11:89–97.

Pickstone, J. V. (1976). Vital actions and organic physics: Henri Dutrochet and French physiology during the 1820s. *Bulletin of the History of Medicine* 50:191–212.

Pilet, P. E. (1962). Jean Senebier, un des précurseurs de Claude Bernard. *Archives Internationales d'Histoire des Sciences* 15:303–313.

Pirie, N. W. (1937). The meaninglessness of the terms life and living. In *Perspectives in Biochemistry* (J. Needham and D. E. Green, eds.), pp. 11–22. Cambridge University Press.

——— (1940). The criteria of purity used in the study of large molecules of biological origin. *Biological Reviews* 15:377–404.

——— (1948). Development of ideas on the nature of viruses. *British Medical Bulletin* 5:329–333.

——— (1954). On making and recognizing life. In *New Biology 16* (M. L. Johnson, M. Abercrombie, and G. E. Fogg, eds.), pp. 41–53. London: Penguin.

——— (1962). Patterns of assumption about large molecules. *Archives of Biochemistry and Biophysics*, Supplement 1:21–29.

——— (1974). Review of S. W. Fox and K. Dose, *Molecular Evolution and the Origin of Life* (1972). *Science Progress* 61:162–163.

Plantefol, L. (1968). Le genre du mot enzyme. *Comptes Rendus de l'Académie des Sciences* 266C: 41–46.

Pliny (1938). *Natural History* (translated by H. Rackham). London: Heinemann.

Pollock, M. R. (1970). The discovery of DNA: An ironic tale of chance, prejudice and insight. *Journal of General Microbiology* 63:1–20.

——— (1976). From pangens to polynucleotides: The evolution of ideas on the mechanism of biological replication. *Perspectives in Biology and Medicine* 19:455–472.

Pontremoli, S., and Melloni, E. (1986). Extralysosomal protein degradation. *Annual Review of Biochemistry* 56:455–481.

Popper, K. R. (1959). *The Logic of Discovery*. London: Hutchinson.

——— (1962). *Conjectures and Refutations*. London: Routledge and Kegan Paul.

——— (1972). *Objective Knowledge*. Oxford University Press.

Prestwich, G. D., and Blomquist, G. J. (eds.) (1987). *Pheromone Biochemistry*. New York: Academic Press.

Przibram, H. (1926). *Die anorganische Grenzgebiete der Biologie*. Berlin: Bornträger.

Pyenson, L. (1977). "Who the guys were:" Prosopography in the history of science. *History of Science* 15:155–188.

Qin, Y., and Simon, H. A. (1990). Laboratory replication of scientific discovery process. *Cognitive Science* 14:281–312.

Querner, H. (1972). Probleme der Biologie um 1900 auf der Versammlung der Deutschen Naturforscher und Aerzte. In *Wege der Naturforschung 1822–1972* (H. Querner and H. Schipperges. eds.), pp. 186–202. Berlin: Springer.

Quine, W. V. O. (1960). *Word and Object*. Cambridge, Mass.: The MIT Press.

Quiocho, F. A. (1990). Atomic structures of periplasmic binding proteins and the high-affinity active transport systems in bacteria. *Philosophical Transactions of the Royal Society* B326:341–351.

Racker, E. (1976). *A New Look at Mechanisms in Bioenergetics*. New York: Academic Press.

Rahn, O. (1932). *Physiology of Bacteria*. Philadelphia: Blakiston.

Randall, J. T. (1951). An experiment in biophysics. *Proceedings of the Royal Society* A208:1–24.

——— (1975). Emmeline Jean Hanson. *Biographical Memoirs of Fellows of the Royal Society* 21:313–344.

Rapkine, L. (1938). Role des groupements sulfhydrilés dans l'activité de l'oxyréductase de triose phosphate. *Comptes Rendus de l'Académie des Sciences* 207:301–304.

Ravetz, J. R. (1971). *Scientific Knowledge and its Social Problems*. Oxford University Press.

Ravin, A. W. (1977). The gene as catalyst: The gene as organism. *Studies in the History of Biology* 1:1–45.

Redhead, M. (1990). Quantum theory. In *Companion to the History of Modern Science* (R. C. Olby et al., eds.), pp. 458–478. London: Routledge.

Rehbock, P. F. (1987). Prosopography? *Ideas and Production* 6:116–121.

Reich, E., Rifkin, D. B. and Shaw, E. (eds.) (1975). *Proteases and Biological Control*. Cold Spring Harbor Laboratories.

Reichert, E. T. and Brown, A. P. (1909). *The Differentiation and Specificity of Corresponding Proteins and other Vital Substances in Relation to Biological Classification and Organic Evolution: The Crystallography of Hemoglobins*. Washington, D.C.: Carnegie Institution.

Reingold, N. (1981). Science, scientists and historians of science. *History of Science* 19:274–283.

——— (1986). History of science today. 1. Uniformity as hidden diversity: History of science in the United States 1920–1940. *British Journal of the History of Science* 19:243–262.

Resnick, D. B. (1989). Adaptationist explanations. *Studies in the History and Philosophy of Science* 20:193–213.

Richardson, J. S. (1981). The anatomy and taxonomy of protein structures. *Advances in Protein Chemistry* 34:167–270.

Rivett, A. J. (1989). High molecular mass intracellular proteinases. *Biochemical Journal* 263: 625–533.

Roberts, H. F. (1929). *Plant Hybridization before Mendel.* Princeton University Press.

Roberts, J. (1990). Pauling: charismatic, controversial and tough. *Chemical and Engineering News* 68(5):68–69.

Robertson, J. M. and Woodward, I. (1937). An X-ray study of phthalocyanins. Part III. Quantitative structure determination of metal phthalocyanine. *Journal of the Chemical Society,* pp. 219–230.

Robinson, G. (1979). *A Prelude to Genetics.* Lawrence, Kans.: Coronado Press.

Rocke, A. J. (1981). Kekulé, Butlerov, and the historiography of the theory of chemical structure. *British Journal for the History of Science* 14:27–57.

——— (1984). *Chemical Atomism in the Nineteenth Century.* Columbus: Ohio State University Press.

——— (1985). Hypothesis and experiment in the early development of Kekulé's benzene theory. *Annals of Science* 42:355–381.

Roe, A. (1972). Patterns in productivity of scientists. *Science* 176:940–941.

Roger, J. (1963). *Les Sciences de la Vie dans la Pensée Francaise du XVIIIe Siècle.* Paris: Armand Colin.

Roll-Hansen, N. (1976). Critical teleology: Immanuel Kant and Claude Bernard on the limitations of experimental biology. *Journal of the History of Biology* 9:59–91.

——— (1978). Drosophila genetics: a reductionist research program. *Journal of the History of Biology* 11:159–210.

——— (1979). Reductionism in biological research. Reflections on some case studies in experimental biology. In *Perspectives in Metascience* (J. Bärmark, ed.), pp. 157–172. Göteborg: Kungl. Vetenskaps- och Vitterhets-Samhället.

Rose, S. (1988). Reflections on reductionism. *Trends in Biochemical Sciences* 13:160–162.

Rosenberg, A. (1985). *The Structure of Biological Science.* Cambridge University Press.

——— (1989). From reductionism to instrumentalism? In *What the Philosophy of Biology Is* (M. Ruse, ed.), pp. 245–262. Dordrecht: Kluwer.

Rosenberg, C. E. (1976). *No Other Gods*. Baltimore: The Johns Hopkins University Press.

Rothschuh, K. E. (1973). *History of Physiology* (translated by G. B. Risse). Huntington, N.Y.: Krieger.

——— (1976). Die Bedeutung apparativer Hilfsmittel für die Entwicklung der biologischen Wissenschaften im 19. Jahrhundert. In *Naturwissenschaft, Technik und Wirtschaft im 19. Jahrhundert* (W. Treue and K. Manuel, eds.), pp. 161–185. Gottingen: Vandenhoeck & Rupprecht.

——— (1980). Medicina historica. Zum Selbstverständniss der historischen Medizin. *Janus* 67:7–19.

Rouvray, D. H. (1975). The pioneers of isomer enumeration. *Endeavour* 35:28–33.

Rubin, L. (1980). Styles of scientific speculation: Paul Ehrlich and Svante Arrhenius on immunochemistry. *Journal of the History of Medicine* 35:397–425.

Rudolph, G. (1983). Das Mechanismusproblem in der Physiologie des 19. Jahrhunderts. *Berichte zur Wissenschaftsgeschichte* 6:7–28.

Rundle, R. E., and French, D. (1943). The configuration of starch in the starch-iodine complex. III. X-ray diffraction studies of the starch-iodine complex. *Journal of the American Chemical Society* 65:1707–1710.

Ruse, M. (1973). *The Philosophy of Biology*. London: Hutchinson.

——— (1987). Biological species: Natural kinds, individuals, or what? *British Journal for the Philosophy of Science* 38:225–242.

Ruse, M. (ed.) (1989). *What the Philosophy of Biology Is*. Dordrecht: Kluwer.

Russell, B. (1959). *My Philosophical Development*. London: Allen & Unwin.

Russell, C. A. (1971). *The History of Valency*. Leicester University Press.

——— (1984). Whigs and professionals. *Nature* 308:777–778.

——— (1988). "Rude and disgraceful beginnings": A view of history of chemistry from the nineteenth century. *British Journal for the History of Science* 21:273–294.

Russell, N. (1988a). Towards a history of biology in the twentieth century: Directed autobiographies as historical sources. *British Journal for the History of Science* 21:77–89.

——— (1988b). Oswald Avery and the origin of molecular biology. *British Journal for the History of Science* 21:393–400.

Ryan, F. J. (1952). Adaptation to use lactose by *Escherichia coli*. *Journal of General Microbiology* 7:69–88.

Ryle, A. P., Sanger, F., Smith, L. F., and Kitai, R. (1955). The disulfide bonds of insulin. *Biochemical Journal* 60:541–556.

Ryle, G. (1954). *Dilemmas*. Cambridge University Press.

Sajet, H. (1978). *L'Essor de la Biologie Moléculaire*. Paris: Centre National de la Recherche Scientifique.

Sandler, I. (1979). Some reflections on the protean nature of the scientific precursor. *History of Science* 17:170–190.

Sanger, F. (1952). The arrangement of amino acids in proteins. *Advances in Protein Chemistry* 7:1–67.

——— (1988). Sequences, sequences, sequences. *Annual Review of Biochemistry* 57:1–28.

Sanger, F., Nicklen, S., and Coulson, R. (1977). DNA sequencing with chain-terminating inhibitors. *Proceedings of the National Academy of Sciences* 74:5463–5467.

Sapp, J. (1987). *Beyond the Gene*. Oxford University Press.

Sarton, G. (1938). In *Cooperation in Research* (Publication No.501), pp. 465–481. Washington, D.C.: Carnegie Institution.

——— *A Guide to the History of Science*. Waltham, Mass.: Chronica Botanica.

Sasso, J. (1980). The stages of the creative process. *Proceedings of the American Philosophical Society* 124:119–132.

Schachman, H. K. (1988). Can a simple model account for the allosteric transition of aspartate transcarbamoylase? *Journal of Biological Chemistry* 263:18583–18586.

Schaffner, K. F. (1974). The peripherality of reductionism in the development of modern biology. *Journal of the History of Biology* 7:111–139.

Schelenz, H. (1911). *Zur Geschichte der pharmazeutisch-chemischen Destilliergeräte*. Berlin: Springer.

Schiller, J. (1967). *Claude Bernard et les Problèmes Scientifiques de son Temps*. Paris: Editions du Cedre.

——— (1980). *Physiology and Classification*. Paris: Maloine.

Schiller, J., and Schiller, T. (1975). *Henri Dutrochet: Le Matérialisme Mécaniste et la Physiologie Générale*. Paris: Blanchard.

Schimmel, P. (1990). Hazards and their exploitation in the applications of molecular biology to structure-function relationships. *Biochemistry* 29:9495–9502.

Schleiden, M. (1849–1850). *Grundzüge der wissenschaftlichen Botanik*. Third ed. Leipzig: Engelmann.

Schlesinger, G. (1963). *Method in the Physical Sciences*. New York: Humanities Press.

Schmidt-Nielsen, K. (1967). The unusual animal, or to expect the unexpected. *Federation Proceedings* 26:981–986.

Schnelle, T., Porus, V. N., Symotnik, S., and Kielanowski, T. (1983). Looking again at Ludwik Fleck. *Kwartalnik Historii Nauki i Techniki* 28:525–587.

Schorer, M. (1961). *Sinclair Lewis: An American Life*. New York: McGraw-Hill.

Schorlemmer, C. (1894). *The Rise and Development of Organic Chemistry*. Revised ed. London: Macmillan.

Schrödinger, E. (1944). *What is Life? The Physical Basis of the Living Cell*. Cambridge University Press.

Schwann, T. (1836). Ueber das Wesen des Verdauungsprocesses. *Archiv für Anatomie, Physiologie und wissenschaftliche Medizin*, pp. 90–138.

——— (1839). *Mikroskopische Untersuchungen über die Übereinstimmung in der Struktur und dem Wachstum der Tiere und Pflanzen*. Berlin: Sander.

Scriven, M. (1959). Explanation and prediction in evolutionary theory. *Science* 130:477–482.

Semenza, G. (ed.) (1981, 1982). *Of Oxygen, Fuels, and Living Matter*. Chichester, Eng.: Wiley.

——— (1983, 1986). *Selected Topics in the History of Biochemistry*. Amsterdam: Elsevier.

Senebier, J. (1802). *Essai sur l'Art d'Observer et de Faire des Expériences*. Geneva: Paschoud.

Sengupta, I. N. (1973). Growth of the biochemical literature. *Nature* 244:75–76,118.

Senior, A. E. (1988). ATP synthesis by oxidative phosphorylation. *Physiological Reviews* 68:177–231.

Serafini, A. (1989). *Linus Pauling: A Man and His Science*. New York: Paragon House.

Servos, J. W. (1983). Review of R. E. Kohler, *From Medical Chemistry to Biochemistry*. *Isis* 74:273–275.

——— (1990). *Physical Chemistry from Ostwald to Pauling*. Princeton University Press.

Shannon, C. E. and Weaver, W. (1949). *The Mathematical Theory of Communication*. Urbana: University of Illinois Press.

Shapere, D. (1984). *Reason and the Search for Knowledge*. Dordrecht: Reidel.

——— (1987). Method in the philosophy of science and epistemology. In *The Process of Science* (N. J. Nersessian, ed.), pp. 1–39. Dordrecht: Nijhoff.

Shapin, S. (1989). The invisible technician. *American Scientist* 77:554–563.

Shapin, S., and Thackray, A. (1974). Prosopography as a research tool in history of science: The British scientific community. *History of Science* 12:1–28.

Shedlovsky, T. (1943). Criteria of purity of proteins. *Annals of the New York Academy of Sciences* 43:259–272.

Shippen-Lentz, D., and Blackburn, E. H. (1990). Functional evidence for an RNA template in telomerase. *Science* 247:546–552.

Sibatani, A. (1981). Two faces of molecular biology: Revolution and normal science. *Rivista di Biologia* 74:279–296.

Siegfried, M. (1916). *Über partielle Eiweisshydrolyse*. Berlin: Bornträger.

Sigman, D. S., and Brazier, M. A. B. (eds.) (1980). *The Evolution of Protein Structure and Function*. New York: Academic Press.

Silverstein, A. M. (1982). Development of the concept of immunological specificity. *Cellular Immunology* 67:396–409; 71:183–195.

——— (1989). *A History of Immunology.* San Diego: Academic Press.

Simmer, H. H. (1978). Die Entdeckung und die Entdecker des Sekretins. *Medizinische Welt* 29:1991–1996.

Simon, M. A. (1971). *The Matter of Life.* New Haven: Yale University Press.

Simpson, G. G. (1964a). Organisms and molecules in evolution. *Science* 146:1535–1538.

——— (1964b). *This View of Life.* New York: Harcourt Brace & World.

Singer, M., and Berg, P. (1990). *Genes and Genomes.* Mills Valley, Calif.: University Science Books.

Sinsheimer, R. L. (1957). First steps toward a genetic chemistry. *Science* 125:1123–1128.

Small, H. G. (1978). Cited documents as concept symbols. *Social Studies of Science* 8:327–340.

Smeaton, W. A. (1963). Guyton de Morveau and chemical affinity. *Ambix* 11:55–64.

——— (1970). Thorbern Olof Bergman. *Dictionary of Scientific Biography* 2:4–8. New York: Charles Scribner's Sons.

——— (1972). Louis Bernard Guyton de Morveau. *Dictionary of Scientific Biography* 5:600–604. New York: Charles Scribner's Sons.

Smith, W. C. (1939). Arthur Hutchinson. *Obituary Notices of Fellows of the Royal Society* 2:483–491.

Snelders, H. A. M. (1984). J. H. van't Hoff's research school in Amsterdam, 1877–1895. *Janus* 71:1–30.

Snow, C. P. (1934). *The Search.* London: Gollancz.

——— (1959). *The Two Cultures and the Scientific Revolution.* Cambridge University Press.

——— (1960). *The Affair.* London: Macmillan.

Snyder, S. H., Sklar, P. B., and Pevsner, J. (1988). Molecular mechanisms of olfaction. *Journal of Biological Chemistry* 263:13971–13974.

Sober, E. (1975). *Simplicity.* Oxford University Press.

Sonneborn, T. M. (1938). Mating types in *Paramecium aurelia. Proceedings of the American Philosophical Society* 79:411–434.

Sørensen, S. P. L. (1930). Die Konstitution der löslichen Proteinstoffe als reversibel dissoziable Komponentensysteme. *Kolloid-Zeitschrift* 53:102–124, 170–199, 306–318.

Stadler, L. J. (1954). The gene. *Science* 120:811–819.

Stanley, W. M. (1935). Isolation of a crystalline protein possessing the properties of tobacco-mosaic virus. *Science* 81:644–645.

Starling, E. H. (1905). The chemical correlation of the functions of the body. *Lancet* 2:339–341.

Steitz, J. A. (1986). Shaping research in gene expression: Role of the Cambridge Medical Research Council Laboratory of Molecular Biology. *Perspectives in Biology and Medicine* 29:S90-S95.

Stent, G. S. (1968). That was the molecular biology that was. *Science* 160:390–395.

——— (1972). Prematurity and uniqueness in scientific discovery. *Scientific American* 228(12): 84–93.

——— (1979). A thought collective. *Quarterly Review of Biology* 54:421–427.

——— (1982). Max Delbrück. *Genetics* 101:1–16.

——— (1989). Light and life: Niels Bohr's legacy to contemporary biology. *Génome* 31:11–15.

Stephenson, M. (1949). *Bacterial Metabolism*. Third ed. London: Longmans Green.

Stern, C., and Sherwood, E. (eds.) (1966). *The Origin of Genetics: A Mendel Source Book*. San Francisco: Freeman.

Stokes, T. D. (1982). The double helix and the warped zipper—An exemplary tale. *Social Studies of Science* 12:207–240.

Stone, L. (1971). Prosopography. *Daedalus* 100:46–79.

Štrbáňová, S. (1981). Biochemical journals and their profile in 1840–1930. *Acta Historiae Rerum Naturalium necnon Technicarum* 16:149–195.

Stryer, L. (1988). *Biochemistry*. Third ed. New York: Freeman.

Stubbe, H. (1965). *Kurze Geschichte der Genetik bis zur Wiederentdeckung der Vererbungsregeln Gregor Mendels*. Second ed. Jena: Fischer.

Sturtevant, A. H. (1959). Thomas Hunt Morgan. *Biographical Memoirs of the National Academy of Sciences* 33:283–325.

Svedberg, T. (1929). Mass and size of protein molecules. *Nature* 123:871.

——— (1937). The ultracentrifuge and the study of high-molecular compounds. *Nature* 139:1051–1062.

Symonds, N. (1986). What is Life? Schrödinger's influence on biology. *Quarterly Review of Biology* 61:221–226.

——— (1988). Schrödinger and Delbrück: Their status in biology. *Trends in Biochemical Sciences* 13:232–234.

Synge, R. L. M., and Williams, E. F. (1990). Albert Charles Chibnall. *Biographical Memoirs of Fellows of the Royal Society* 35:57–96.

Tanford, C. (1980). *The Hydrophobic Effect*. Second ed. New York: Wiley.

——— (1989). *Ben Franklin Stilled the Waves*. Durham, N.C.: Duke University Press.

Taylor, C. W., and Barron, F. (eds.) (1963). *Scientific Creativity, Its Recognition and Development*. New York: Wiley.

Taylor, F. S. (1953). The idea of the quintessence. In *Science, Medicine, and History* (E. A. Underwood, ed.), vol. 1, pp. 247–265. Oxford University Press.

Teich, M. (1973). From "enchyme" to "cytoskeleton." In *Changing Perspectives in the History of Science* (M. Teich and R. Young, eds.), pp. 439–471. London: Heinemann.

——— (1975). A single path to the double helix? *History of Science* 13:264–283.

——— (1980). A history of biochemistry. *History of Science* 18:46–67.

——— (1981). Ferment or enzyme: What's in a name? *History and Philosophy of the Life Sciences* 3:193–215.

Temkin, O. (1946). Materialism in French and German physiology in the early nineteenth century. *Bulletin of the History of Medicine* 20:322–327.

Thackray, A. W. (1966). The chemistry of history. *History of Science* 5:124–134.

——— (1980). The pre-history of an academic discipline: The study of the history of science in the United States 1891–1941. *Minerva* 18:448–473.

Thackray, A. W., and Merton, R. K. (1972). On discipline building: The paradoxes of George Sarton. *Isis* 63:473–495.

Thackray, A. W., Sturchio, J. L, Carroll, P. T., and Bud, R. (1985). *Chemistry in America, 1876–1976.* Dordrecht: Reidel.

Timoféeff-Ressovsky, N. W., Zimmer, H. G., and Delbrück, M. (1935). Uber die Natur der Genmutation und der Genstruktur. *Nachrichten von der Gesellschaft für Wissenschaften zu Göttingen, Math.-Phys. Klasse, Fachgruupe VI,* pp. 189–245.

Todd, A. (1983). *A Time to Remember.* Cambridge University Press.

Tondl, L. (1979). Science as a vocation. In *Perspectives in Metascience* (J. Bärmark, ed.), pp. 173–184. Göteborg: Kungl. Vetenskaps- och Vitterhets-Samhället.

Toulmin, S. (1967). The evolutionary development of natural science. *American Scientist* 55:456–471.

Tristram, G. R. (1949). Amino acid composition of purified proteins. *Advances in Protein Chemistry* 5:83–153.

Troland, L. T. (1917). Biological enigmas and the theory of enzyme action. *American Naturalist* 51:321–350.

Turner, R. S. (1982). Justus Liebig versus Prussian chemistry. Reflections on early institute building in Germany. *Historical Studies in the Physical Sciences* 13:129–162.

Turner, S., Kerwin, E., and Woolwine, D. (1984). Careers and creativity in nineteenth-century physiology: Zloczower *redux. Isis* 75:523–529.

Umbarger, H. E. (1956). Evidence for a negative-feedback mechanism in the biosynthesis of isoleucine. *Science* 123:848.

Uschmann, G. (1961). Ernst Haeckel. Third ed. Leipzig: Urania.

Vane, J., and Cuatrecasas, P. (1984). Genetic engineering and pharmaceuticals. *Nature* 312:303–305.

Van Niel, C. B. (1957). Albert Jan Kluyver. *Journal of General Microbiology* 16:499–521.

Verkade, P. E. (1985). *A History of the Nomenclature of Organic Chemistry.* Dordrecht: Reidel.

Verworn, M. (1903). *Die Biogen Hypothese.* Jena: Fischer.

Vesterberg, O. (1989). History of electrophoretic methods. *Journal of Chromatography* 480:3–19

Waddington, C. H. (1969). Some European contributions to the prehistory of molecular biology. *Nature* 221:318–321.

Wade, N. (1981). *The Nobel Duel*. Garden City, N.Y.: Doubleday.

Waksman, S. A. (1953). *Sergei N. Winogradsky, His Life and Work*. New Brunswick, N.J.: Rutgers University Press.

Walden, P. (1935). Nationale Wege der modernen Chemie. *Chemiker-Zeitung* 59:2–3.

——— (1941). *Geschichte der organischen Chemie seit 1880*. Berlin: Springer.

Walker, J. E., Fearnley, I. M., Lutter, R., Todd, R. J., and Runswick, M. J. (1990). Structural aspects of proton-pumping ATP-ases. *Philosophical Transactions of the Royal Society* B326:367–378.

Wallace, D. B., and Gruber, H. E. (eds.) (1989). *Creative People at Work*. Oxford University Press.

Walsh, C. (1979). *Enzymatic Reaction Mechanisms*. San Francisco: Freeman.

Warburg, O. (1929). Atmungsferment und Oxydasen. *Biochemische Zeitschrift* 214:1–3.

——— (1938). Chemische Konstitution von Fermenten. *Ergebnisse der Enzymforschung* 7:210–245.

——— (1948). *Wasserstoffübertragende Fermente*. Berlin: Saenger.

Warner, J. H. (1985). Science in medicine. *Osiris* [2] 1:37–58.

Wassermann, A. (1910). Ueber den Einfluss des Spezifitätsbegriffs auf die moderne Medizin. *Deutsche medizinische Wochenschrift* 36:1860–1863.

Waterson, A. P., and Wilkinson, L. (1978). *An Introduction to the History of Virology*. Cambridge University Press.

Watson, J. D. (1968). *The Double Helix*. New York: Atheneum.

——— (1980). *The Double Helix* (G. Stent, ed.). New York: Norton.

Weaver, W. (1970a). Molecular biology: Origin of the term. *Science* 152:581–582.

——— (1970b). *Scene of Change*. New York: Charles Scribner's Sons.

Webb, E. C. (1970). Communication in biochemistry. *Nature* 225:132–135.

——— (1984). *Enzyme Nomenclature 1984: Recommendations of the Nomenclature Committee of the International Union of Biochemistry on the Nomenclature and Classification of Enzyme-Catalyzed Reactions*. Orlando, Fla.: Academic Press.

Weber, G. (1990). Whither biophysics? *Annual Review of Biophysics and Bio-physical Chemistry* 19:1–6.

Weiss, P. (1947). The problem of specificity in growth and development. *Yale Journal of Biology and Medicine* 19:235–278.

Wells, R. D. (1988). Unusual DNA structures. *Journal of Biological Chemistry* 263:1095–1098.

Westheimer, F. H. (1985). The discovery of the mechanisms of enzyme action 1947–1963. *Advances in Physical Organic Chemistry* 21:1–34.

——— (1986). Polyribonucleic acids as enzymes. *Nature* 319:534–536.

Weyer, J. (1973). Chemiegeschichtsbeschreibung im 19. und 20. Jahrhundert. *Sudhoffs Archiv* 57:171–194.

——— (1974). *Chemiebeschreibung von Wiegleb (1790) bis Partington (1970)*. Hildesheim: Gerstenberg.

Whitehead, A. N. (1925). *Science and the Modern World*. New York: Macmillan.

Whitt, L. A. (1990). Atoms or affinities? The ambivalent reception of the Daltonian theory. *Studies in the History and Philosophy of Science* 21:57–89.

Whittaker, R. H. (1969). New concepts of kingdoms of organisms. *Science* 163:150–160.

Wiener, N. (1948). *Cybernetics, or the Control and Communication in the Animal and the Machine*. New York: Wiley.

Wilkie, J. S. (1960–1961). Nägeli's work on the fine structure of living matter. *Annals of Science* 16:11–42, 171–207, 209–239; 17:27–62.

Wilkins, M. H. F. (1987). John Turton Randall. *Biographical Memoirs of Fellows of the Royal Society* 33:493–535

Wilkinson, L. (1976). The development of the virus concept as reflected in corpora of studies on individual pathogens. 3. Lessons of the plant viruses—Tobacco mosaic virus. *Medical History* 20:111–134.

Williams, L. P. (1965). *Michael Faraday, a Biography*. London: Chapman and Hall.

——— (1966). The historiography of Victorian science. *Victorian Studies* 9:197–204.

Williams, R. (1985). *Keywords*. Second ed. Crook Helm: Fontana.

Williams, R. J. P. (1987). The functions of structure and dynamics in proteins, peptides, and metal ion complexes and their relationship to biological recognition and the handling of information. *Carlsberg Laboratory Communications* 52:1–30.

Wilson, E. B. (1895). *An Atlas of the Fertilization and Karyokinesis of the Ovum*. New York: Macmillan.

——— (1896). *The Cell in Development and Inheritance*. New York: Macmillan.

——— (1923). The physical basis of life. *Science* 57:277–286.

——— (1925). *The Cell in Development and Heredity*. New York: Macmillan.

Wilson, H. R. (1988). The double helix and all that. *Trends in Biochemical Sciences* 13:275–278

Wilson, L. G. (1980). Medical history without medicine. *Journal of the History of Medicine* 35:5–7.

——— (1983). Review of R. E. Kohler *From Medical Chemistry to Biochemistry*. *Journal of the History of Medicine* 38:462–464.

Witkowski, J. A. (1979). Alexis Carrel and the mysticism of tissue culture. *Medical History* 23:276–296.

——— (1980a). Dr. Carrel's immortal cells. *Medical History* 24:129–142.

——— (1980b). W. T. Astbury and Ross G. Harrison: The search for the molecular determination of form in the developing embryo. *Notes and Records of the Royal Society* 35:195–219.

——— (1985). The magic of numbers. *Trends in Biochemical Sciences* 10:139–141.

——— (1987). Optimistic analysis—Chemical embryology in Cambridge, 1920–1942. *Medical History* 31:247–268

——— (1990). Carrel's cultures. *Science* 247:1385–1386.

Wittgenstein, L. (1922). *Tractatus Logico-Philosophicus*. London: Routledge and Kegan Paul.

Woese, C. R., Kandler, O., and Wheelis, M. L. (1990). Toward a natural system of organisms: Proposal for the domains Archaea, Bacteria, and Eucaria. *Proceedings of the National Academy of Sciences* 87:4576–4579.

Wolfenden, R. (1976). Transition-state analog inhibitors and enzyme catalysis. *Annual Review of Biophysics and Bioengineering* 5:271–306.

Wotiz, J. H., and Rudofsky, S. (1987). The unknown Kekulé. In *Essays on the History of Organic Chemistry* (J. G. Traynham, ed.), pp. 21–34. Baton Rouge: Louisiana State University Press.

Wright, R. D. (1978). The origin of the word "hormone." *Trends in Biochemical Sciences* 3:275.

Wright, S. (1917). Color inheritance in animals. *Journal of Heredity* 8:224–235.

——— (1945). Physiological aspects of genetics. *Annual Review of Physiology* 7:75–106.

——— (1986). Recombinant DNA technology and its social transformation. *Osiris* 2(2): 303–360.

Wrinch, D. (1938). On the hydration and denaturation of proteins. *Philosophical Magazine* 25:705–739. On the molecular weights of globular proteins. Ibid. 26:313–332.

Wurtz, A. (1869). *Histoire des Doctrines Chimiques depuis Lavoisier jusqu'à nos Jours*. Paris: Hachette.

Wüthrich, K. (1990). Protein structure determination in solution by NMR spectroscopy. *Journal of Biological Chemistry* 265:22059–22062

Wyatt, H. V. (1974). How history has blended. *Nature* 249:803–805.

——— (1975). Knowledge and prematurity: The journey from transformation to DNA. *Perspectives in Biology and Medicine* 18:149–156.

Wyman, J. (1948). Heme proteins. *Advances in Protein Chemistry* 4:407–531.

——— (1949). Some physico-chemical evidence regarding the structure of haemoglobin. In *Haemoglobin* (F. J. W. Roughton and J. C. Kendrew, eds.), pp. 95–106. London: Butterworths.

——— (1964). Linked functions and reciprocal effects in hemoglobin: a second look. *Advances in Protein Chemistry* 19:223–286.

——— (1972). On allosteric models. *Current Topics in Cellular Regulation* 6:209–226.

Wyman, J., and Allen, D. W. (1951). Heme interactions in hemoglobin and the basis of the Bohr effect. *Journal of Polymer Science* 7:499–518.

Yates, R. A. and Pardee, A. B. (1956). Control of pyrimidine biosynthesis in Escherichia coli by a feed-back mechanism. *Journal of Biological Chemistry* 221:767–770.

Yoxen, E. J. (1979). Where does Schroedinger's "What is Life?" belong in the history of molecular biology? *History of Science* 17:17–52.

——— (1982). Giving life a new meaning: The rise of the molecular biological establishment. In *Scientific Establishments and Hierarchies* (N. Elias, H. Martins, and R. Whitley, eds.), pp. 123–143. Dordrecht: Reidel.

Zaug, A. J., and Cech, T. R. (1986). The intervening sequence RNA of *Tetrahymena* is an enzyme. *Science* 231:470–475.

Ziman, J. (1981). What are the options? Social determinants of personal research plans. *Minerva* 19:1–42.

——— (1984). *An Introduction to Science Studies.* Cambridge University Press.

Zuckerman, H. (1967). Nobel laureates in science: Patterns of productivity, collaboration and authorship. *American Sociological Review* 32:391–403.

——— (1977). *Scientific Elite.* New York: Free Press.

Zuckerman, H., and Merton, R. K. (1971). Patterns of evaluation in science: Institutionalisation, structure, and functions of the referee system. *Minerva* 9:66–100.

Zuckerman, J. J. (1987). The chemist as teacher of history. *Journal of Chemical Education* 64:828–835.

Zumbach, C. (1984). *The Transcendent Science: Kant's Conception of Biological Methodology.* The Hague: Nijhoff.

Index of Personal Names

Index of Subjects

Photo by Barrie-Kent

Joseph S. Fruton is Eugene Higgins Professor, Emeritus, of Biochemistry at Yale University and a distinguished scholar whose writings on the field have been widely praised. His recent book, *Contrasts in Scientific Style: Research Groups in the Chemical and Biochemical Sciences*, published by the American Philosophical Society in 1990, received the John Frederick Lewis Award of that society.

HARVARD UNIVERSITY PRESS
Cambridge, Massachusetts · London, England

Design by Gwen Frankfeldt